PHOTO-ACTIVATED SLUDGE: A NOVEL ALGAL-BACTERIAL BIOTREATMENT FOR NITROGEN REMOVAL FROM WASTEWATER

ANGÉLICA RADA ARIZA

Thesis committee

Promotor
Prof. Dr Piet N.L. Lens
Professor of Environmental Biotechnology
IHE-Delft Institute for Water Education, the Netherlands

Co-promotors
Dr Peter van der Steen
Associate Professor of Environmental Technology
IHE-Delft Institute for Water Education, the Netherlands

Dr Carlos M. Lopez Vazquez
Associate Professor of Sanitary Engineering
IHE-Delft Institute for Water Education, the Netherlands

Other members

Prof. Dr Oene Oenema, Wageningen University & Research

Prof. Dr Jean-Philippe Steyer, l'Institut National de la Recherche Agronomique-INRA, France

Prof. Dr Diederik P.L. Rousseau, Ghent University, Belgium

Dr Anthony Verschoor, KWR Water Cycle Research Institute, Nieuwegein, The Netherlands

This research was conducted under the auspices of the SENSE Research School for Socio-Economic and Natural Sciences of the Environment

Photo-Activated Sludge: A novel algal-bacterial biotreatment for nitrogen removal from wastewater

Thesis

submitted in fulfilment of the requirements of
the Academic Board of Wageningen University and
the Academic Board of the IHE Delft, Institute for Water Education
for the degree of doctor
to be defended in public
on Friday, 16 November 2018 at 1:30 p.m.
in Delft, the Netherlands

by

Angélica Rada Ariza
Born in Santa Marta, Colombia

CRC Press/Balkema is an imprint of the Taylor & Francis Group, an informa business

Published by:
CRC Press/Balkema
Schipholweg 107C, 2316 XC, Leiden, the Netherlands
Pub.NL@taylorandfrancis.com
www.crcpress.com – www.taylorandfrancis.com
ISBN 978-0-367-17886-4 (T&F)
ISBN 978-94-6343-350-1 (WUR)
DOI: https://doi.org/10.18174/459979

To my loving family,

Acknowledgements

This thesis has been one of the most rewarding achievements in my professional life. During this process, I have met amazing, intelligent, caring and helpful people that helped me and supported me to produce this thesis. I would like to express my gratitude to my professor Dr. Piet Lens for his support and encouragement, for his time and all the scientific conversations that guided me and always directed me in the right path.

Equally important, I want to express my sincere gratitude to my mentors Dr. Peter van der Steen and Dr. Carlos Lopez Vazquez. You both taught me to not just to understand more my topic, but to understand and learn how to do research. I want to thank you also for your critical assessment, for always questioning me, sometimes it was not easy, but I can say with certainty that it helped me greatly to build up my confidence in the subject. Your incredible and infinite inputs made this thesis stronger in every way. Furthermore, you were also very supportive in not just the scientific aspects of this thesis, but also in other personal aspects. Thank you very much!.

I want to acknowledge my financial sponsor: Colciencias. Thanks for awarding me with a scholarship to pursue my PhD research, and thanks for investing in the education of the country. Hopefully, this will be reflected in many positive ways, and not just to grow as a country, but to grow as society. I will be honour to be part of this change.

I want also to thank to all the staff in the laboratory: Fred, Frank, Lyzzette, Ferdi, Berend and Peter, it was not always easy to do the experiments, but it was certainly easier with you. Thank you for your help and for distracting me in the long hours with your conversations and cheerings!. Also I want to thank Jolanda Boots, always very diligent and very helpful when it comes to PhD administration ☺.

Also, I was very lucky to have such good MSc. students to mentor, and from all them I learnt something, and I hope I gave something back: Afrin, Nadya, Andres, Larissa, Puji, and Erin, thank you and I wish nothing but the best in your carers.

I want to thank all the people who have made my life in Delft more enjoyable and also help me either way with a supportive word, or even with scientific inputs: Assiyeh, Saul, Diego, Andreita, Gonzalo, Angela, Rohan, Leah, Arlex, Natasha, Leonardo, Marianne, Anika, Adriana and Thijmen, Migue and Pin. Furthemore, special thanks to an amazing designer Marvin Stiefelhagen, without you the picture of my cover would have been like a kid's drawing or even worse.

To my friends in the distance: Ana Elena, Pochy, Lina, Lore, Anita, Leonardo O., Armando, Adri, Paloma, Benly, Carola and Joanina no matter how far, I always felt you were here. No matter how often we talked, or how often we updated each other, you were always there, and will be always there. Every time we meet feels like we have never been apart. Thank youuu!!!. Lots of love for you guys!.

Special thanks to Erika, Pato, Alida, Juliette, Vero, Neiler, Aki, Yared, Juanca, Pablo, and Juan Pa and also the new generation Sophie and Emi ☺. Jessy, Mauri and Larita ☺, thanks for coming all the way down to the Netherlands to share this special moment with me, to Mau thanks for the thesis template, it made my life easier. You are all great friends, always there, and always cheerful, thank you all for your friendship, I hope that the distance or time never let us forget the memories we treasure, and that life give us opportunities to make new ones in the future.

Also to a great group of friends Peter, Berend (yes Berend you are double), Mohaned, and Jeffrey thank you for being there, for always asking, for always remind me that I can, for your advices, your laughs, and even more thanks for all the moments we have shared, and we will continue sharing.

To Nata and Fer you are such a extraordinary women, and I am lucky and proud to call you my friends, you both are one of a kind. I have learnt from each of you so much, you have never let me down, and you have became a rock while here in Delft. You are like sisters to me. Let's cheer for more adventures, anecdotes, talks, learning experiences and stories!. Las quiero muchoo!!.

A mi Familia no hay palabras para expresar el amor, el agradecimiento y admiración que siento por ellos. Ellos son las personas más valiosas e importantes de mi vida: Mi papá, mamá, Alicia, Alexandra, Anita, Toto, Maria Laura e Isabella. Nobles, honestos, responsables, amorosos, caritativos y mucho más, gracias a ustedes hoy soy lo que soy, y todo lo que llegaré a ser. Son las mejores personas que conozco!. Esta tesis es para ustedes, los amo con todo mi corazón. No podía dejar de mencionar a mi tio Luis y Edilma, mi tio Dario y Meche, y a mis primos (que si los escribo todos, no alcanzo), los quiero mucho y siempre me siento en casa cuando estoy con ustedes. Also, to my second family Nurdan, Osman, Nilay, Erik, Tim and Lisa, you are so lovable and kind, thanks for your care and love. Love you.

Last but not the least to Can, thank you for your support, you are a source of inspiration, kindness and love. You are the most patient, loving, caring, selfless, intelligent and amazing man I have ever met. I am grateful to have your love, and even more thankful you came into my life, this is also your thesis in a way. Te amo.

Thank you all from the bottom of my heart,

Angélica Rada

SUMMARY

Nitrogen rich wastewaters (10-400 mg N L^{-1}) are usually produced by municipal, industrial and agricultural wastes, such as effluents from anaerobic treatments. These represent a risk to the environment due to the high nutrient concentrations (nitrogen and phosphorous), which can cause eutrophication of water bodies, deteriorating the quality of the ecosystems. As a solution, the potential nitrogen removal capacity of a novel bio-treatment system, namely the Photo-Activated Sludge (PAS), composed of microalgae and bacteria consortia, was studied.

Experimental work using photobioreactors for the cultivation of microalgae and bacteria under sequencing batch conditions (Chapters 3, 4 and 5) showed that microalgal-bacterial consortia can remove ammonium 50% faster than solely microalgal consortia (Chapter 3). The increase in ammonium removal rates was due to the action of nitrifying bacteria, supplied with oxygen produced by algae. Also, the addition of bacteria to the microalgal culture increased the biomass retention, which allowed to uncouple the hydraulic retention time (HRT) and solids retention time (SRT) (Chapter 3). In all experiments, nitrification was the main ammonium removal mechanism within the microalgal-bacterial biomass, followed by algal uptake and nutrient requirements for bacterial growth (Chapters 3, 4, 5 and 7). Carbon oxidation and denitrification were the main removal mechanisms for organic carbon (Chapters 4 and 5). Hence, the role of algae within the microalgal-bacterial system is to provide oxygen to support the aerobic processes. The microalgal-bacterial system offers the possibility of reducing the hydraulic retention time, which can decrease the large area requirements often demanded by algal systems (Chapter 3 and 4).

The SRT was identified as the main parameter to control the efficiency of the technology (Chapter 4). The control of the suspended solids concentration, by adjusting the SRT, influences the light penetration within the reactor, which can limit or enhance the oxygen production by algae (growth rate). In Chapters 5 and 6, a mathematical model for microalgal-bacterial systems that can describe the microbiological processes occurring within the microalgal-bacterial consortia was proposed. The results provided by the model identified the light extinction coefficient of the microalgal-bacterial biomass as the most sensitive parameter of the system. Furthermore, the model was used to evaluate certain scenarios and estimate the optimum SRT required for microalgal-bacterial systems, which seems to lie between 5 and 10 days.

Chapter 7 showed, using respirometric tests with microalgal-bacterial biomass, that the main nitrogen removal mechanism is the uptake by algae, where part of the nitrogen is stored within the cell, and part used for growth. Furthermore, the nitrogen storage by algae was introduced in the model, and the process was calibrated using data from the respirometric tests. Thus, the maximum amount of nitrogen stored by algae could be calculated to be 0.33 grams of nitrogen per gram of algal biomass.

This thesis demonstrated that photo-activated sludge systems using microalgal-bacterial consortia are a sustainable treatment option for ammonium rich wastewaters, providing clean effluents and opening reuse options for the biomass. Furthermore, the PAS systems can reduce the area requirements by halve in comparison with algal systems and likely have a positive energy balance, since the sun is one of the main sources of energy.

SAMENVATTING

Ammoniumrijk afvalwater wordt veelal geproduceerd door gemeentelijk, industrieel en landbouwafval, en effluent uit anaerobe afvalwaterzuiveringsmethoden. Dit vormt een risico voor het milieu vanwege de hoge concentratie aan voedingsstoffen (stikstof en fosfor), wat eutrofiëring in waterpartijen kan bevorderen en daarmee de kwaliteit van ecosystemen kan aantasten. Als innovatieve oplossing hierop is een nieuw biologisch verwerkingsmechanisme genaamd Photo-Activated Sludge (PAS) geëvalueerd, wat gebruik maakt van een consortium van microalgen en bacteriën voor de zuivering van ammoniumrijk afvalwater.

Experimenteel onderzoek met fotobioreactoren voor de cultivering van microalgen en bacteriën onder sequentiële batch-condities (Hoofdstuk 3, 4 en 5) toont aan dat microalgen-bacteriële consortia ammonium 50% sneller verwijderen dan pure microalgen consortia (Hoofdstuk 3). De snelheidstoename in ammoniumverwijdering is een gevolg van de activiteit van nitrificeerders, met zuurstof aangeleverd door algen. Bovendien zorgt de toevoeging van bacteriën aan het microalgen consortium voor een toename van biomassa retentie, wat het ontkoppelen van de hydraulische verblijftijd (HRT) en slib verblijftijd (SRT) toestaat (Hoofdstuk 3). Voor alle experimenten geldt dat nitrificatie het voornaamste ammoniumverwijderingsmechanisme is, gevolgd door algenopname en de opname van voedingsstoffen door bacteriën (Hoofdstuk 3, 4, 5 en 7). Koolstofoxidatie middels denitrificatie was het voornaamste verwijderingsmechanisme van organisch koolstof (Hoofdstuk 4 en 5). De rol van algen in een microalgen-bacterieel consortium is om zuurstof te leveren aan alle aërobe processen. Het microalgen-bacteriëel systeem biedt de mogelijkheid om de

HRT te verlagen en daarmee de grote landbehoefte, wat voor algensystemen vaak wordt vereist, te verlagen (Hoofdstuk 3 en 4).

De SRT is geïdentificeerd als zijnde de voornaamste parameter om de efficiëntie van de technologie aan te passen (Hoofdstuk 4). Het aanpassen van de suspensie met behulp van de SRT beïnvloedt de lichtdoorlating in de reactor, waarmee de zuurstofproductie door algen beïnvloed kan worden. In Hoofdstuk 5 en 6 is een wiskundig model van het microalgen-bacteriëel systeem voorgesteld, waarmee microbiologische processen in het microalgen-bacteriëel consortium worden beschreven. Met de resultaten verkregen met het model is de lichtdoorlaatbaarheidscoëfficiënt van de microalgen-bacterieel biomassa als meest gevoelige parameter geïdentificeerd. Bovendien is het model gebruikt om verscheidene scenario's te evalueren en de optimale SRT voor microalgen-bacteriële systemen is gedefiniëerd tussen 5 en 10 dagen.

De laatste bevindingen van Hoofdstuk 7, verkregen met respirometrische testen, resulteerden in de identificatie van het lot van stikstof wanneer het wordt opgenomen door algen, waar een deel van de stikstof is opgeslagen in de cel, en een deel wordt gebruikt voor groei. Verder is aan de hand van gedetailleerde data van de respirometrische testen stikstofopslag door algen toegevoegd aan het model. Hiermee is berekend dat de maximale hoeveelheid opgeslagen stikstof door algen 0.33 gram stikstof per gram algenbiomassa is.

Samenvattend, dit proefschrift demonstreert dat microalgen-bacteriële consortia een duurzame zuiveringsmethode voor ammoniumrijk afvalwater vormen, wat schoon effluent produceert en tevens toepassingen biedt voor het hergebruiken van biomassa en gezuiverde het effluent. Tenslotte kan hiermee de landbehoefte worden gehalveerd, in vergelijking met algensystemen, en heeft deze technologie een positieve energiebalans, omdat de zon de grootste energiebron is.

CONTENTS

Acknowledgements vii

Summary xi

Samenvatting xiii

1 General introduction 1

1.1 Background 2

1.1.1 Wastewater treatment with algal technologies 3

1.1.2 Wastewater treatment using microalgal-bacterial systems 5

1.2 Problem statement 6

1.3 Outline of thesis 9

2 Microalgal-bacterial consortia for wastewater treatment: a review 13

2.1 Microalgal-bacterial consortia 14

2.1.1 Interactions within microalgal-bacterial consortia 16

2.1.2 Nutrient removal by microalgal-bacterial consortia 19

2.1.3 Microalgal-bacterial systems and configurations 24

2.1.4 Limiting and operational conditions of microalgal-bacterial photobioreactors 26

2.2 Microalgal-bacterial modelling 33

2.3 Aims of this PhD research 36

3 Nitrification by microalgal-bacterial consortia for ammonium removal in a flat panel sequencing photobioreactor 39

3.1 Introduction 40

3.2 Materials and methods 42

3.2.1 Reactor set-up ... 42

3.2.2 Inoculation ... 45

3.2.3 Composition of the synthetic wastewater ... 45

3.2.4 Experimental design ... 46

3.2.5 Sampling and analytical methods ... 47

3.2.6 Nitrogen balance ... 47

3.3 Results and discussion ... 48

3.3.1 Biomass concentration and production in the FPRs ... 48

3.3.2 Solids retention time and the effect on ammonium removal rates ... 52

3.3.3 Fate of nitrogen in the FPRs ... 56

3.3.4 Total and specific ammonium removal rates by algae and nitrifiers in the FPRs ... 62

3.3.5 Implications of using microalgal-bacterial consortia for ammonium removal 65

3.4 Conclusions ... 67

4 Ammonium removal mechanisms in a microalgal-bacterial sequencing-batch photobioreactor at different SRT ... 69

4.1 Introduction ... 71

4.2 Materials and methods ... 74

4.2.1 Photobioreactor set-up ... 74

4.2.2 Growth medium, microalgal-bacterial consortia and inoculation ... 76

4.2.3 Sampling and analytical methods ... 78

4.2.4 Biomass productivity, nitrogen and oxygen mass balance equations 79

4.2.5 Total specific and volumetric ammonium removal rate ... 80

4.2.6 Oxygen mass balance ... 81

4.3 Results and Discussion 82
4.3.1 Biomass concentration and chlorophyll-*a* 82
4.3.2 Nitrogen and ammonium removal efficiencies and rates 85
4.3.3 Nitrogen removal mechanisms and biomass characterization 91
4.3.4 Oxygen production in a microalgal-bacterial photobioreactor under different SRTs 96
4.3.5 Effects of SRT on the light penetration, ammonium removal mechanisms and oxygen production 100
4.4 Conclusions 102
5 Modelling of nitrogen removal using a microalgal-bacterial consortium 103
5.1 Introduction 104
5.2 Materials and Methods 107
5.2.1 Experimental 107
5.2.2 Integrated microalgal-bacterial model 112
5.2.3 Statistical analysis 116
5.3 Results and Discussion 116
5.3.1 Experimental 116
5.3.2 Integrated microalgal-bacterial model 123
5.4 Conclusions 127
6 Modelling of nitrogen removal using a microalgal-bacterial consortium under different SRTs 129
6.1 Introduction 131
6.2 Materials and Methods 134
6.2.1 Microalgal-bacterial model 134
6.2.2 Sensitivity analysis 135

6.2.3 Reactor and data collected 136

6.2.4 Calibration and validation of the microalgal-bacterial model 137

6.2.5 Calculation of the error 141

6.2.6 Evaluation of shorter SRTs 141

6.3 Results and discussion 142

6.3.1 Sensitivity analysis 142

6.3.2 Calibration and validation of the N-compounds, oxygen and COD in batch operational mode 147

6.3.3 Calibration and validation of the biomass characterization and production in sequencing batch mode operation 148

6.3.4 Calibration and validation under sequencing batch operational mode of then concentrations of N-compounds, oxygen and organic carbon 152

6.3.5 Growth rate in a microalgal-bacterial consortium 161

6.3.6 SRT optimization using the microalgal-bacterial model 165

6.4 Conclusions 170

7 Respirometric tests for microalgal-bacterial biomass: modelling of nitrogen storage by microalgae 173

7.1 Introduction 174

7.2 Materials and Methods 177

7.2.1 Microalgal-bacterial parent reactor 177

7.2.2 Respirometric test methodology 179

7.2.3 Modelling of nitrogen storage and utilization of stored nitrogen by microalgae 183

7.3 Results and discussion 186

7.3.1 Solids concentration and light attenuation coefficient in the base microalgal-bacterial reactor 186

7.3.2 Ammonium removal rates, efficiency and biomass characterization of the base microalgal-bacterial reactor 189

7.3.3 Nitrogen storage by microalgae in a microalgal-bacterial biomass. 191

7.3.4 Phototrophic growth on stored nitrogen 197

7.3.5 Modelling the nitrogen storage by algae in a microalgal-bacterial biomass 200

7.4 Conclusions 210

8 Conclusions and recommendations 211

8.1 Introduction 212

8.2 Advantages of microalgal-bacterial consortia for ammonium removal.. 215

8.2.1 Advantages on ammonium removal rates 215

8.2.2 Operational conditions and area requirements 217

8.2.3 Photo-oxygenation and algal harvesting 220

8.3 Influence of the SRT on the operation of a microalgal-bacterial photobioreactor 221

8.4 Evaluation of the microalgal-bacterial consortia using mathematical models 223

8.4.1 Mathematical model for analysis of phototrophic growth, nitrification/denitrification and organic carbon removal processes 223

8.4.2 Respirometric tests and mathematical model for the analysis of nitrogen storage by microalgae 226

8.5 Outlook and concluding remarks 227

Appendix A 231

Appendix B 237

Appendix C 253

Appendix D 263

Appendix E 277

References 281

List of acronyms 309

List of Tables 313

List of figures 315

About the author 321

1

General introduction

1.1 Background

The environment has always been affected by anthropogenic activities, reflected in urbanization and industrialization. Water is one of the most important natural resources, and it is vital for the environment and the human population. Wastewater originated from households, industries, and agriculture is one of the main sources of water pollution. Yet, in 2017 up to 80% of the total wastewater generated worldwide is not treated (UNESCO, 2017). This causes environmental problems such as eutrophication, bioaccumulation of toxic compounds and oxygen depletion (UNESCO, 2017). The treatment and reuse of wastewater is becoming imperative due to scarcity and pollution in some areas, and has important benefits such as protection of the environment, reduction of fresh water consumption and thereby intrinsic economic benefits.

Several treatment technologies are being applied around the world for wastewater treatment. They comprise centralized and decentralized systems such as activated sludge systems, upflow anaerobic sludge blanket (UASB) reactors, anaerobic filters, anaerobic baffled reactors, stabilization ponds, wetlands, high rate algae ponds (HRAP), membrane bioreactors (MBR), and soil aquifer treatment (SAT). As concluded by Noyola et al. (2012), based on a survey carried out over 2734 wastewater treatment facilities located in Latin America and the Caribbean (80% of the 2734 facilities analyzed), the most representative technologies in Latin America and the Caribbean are activated sludge systems, waste stabilization ponds, and UASB. Noyola et al. (2012) reported that in terms of energy consumption per cubic meter of treated wastewater (kWh m^{-3}), stabilization ponds have the lowest energy consumption, followed by UASB reactors coupled with activated sludge for nutrient removal, while activated sludge systems have the highest energy consumption, mainly due to the aeration demands of the process. Tandukar et al. (2007) reported an energy consumption for an activated sludge system of 2-3 kW h^{-1} $kgBOD^{-1}$ removed. The high energy consumption in these systems is attributed to aeration, which represents between 54-97% of the total energy requirements of the plant (Young and Koopman, 1991).

Recently, natural methods for the treatment of wastewater streams have gained higher relevance due to their lower energy requirements, reliability and high removal efficiency of organic and inorganic compounds, in addition to their simplicity and lower operational costs (Abdel-Raouf et al., 2012). Among them, algae present photosynthetic capabilities by using the solar energy for cell growth, offering potential valuable biomass while removing diverse pollutants via different removal mechanisms (de la Noue and de Pauw, 1988). Algae are floating unicellular microorganisms, most of them phototrophic, that perform photosynthesis using light as source of energy and H_2O as electron donor. Phototrophic microalgae grow in the presence of different minerals, nutrients and CO_2 as their carbon source (Bitton 2005).

1.1.1 Wastewater treatment with algal technologies

Wastewater treatment using microalgae and microalgal-bacterial consortia can provide clean effluents free of organic and inorganic compounds, heavy metals, and pathogens, and simultaneously produce useful biomass, which can be used for the production of biofuels, fertilizers and other bioproducts (Samorì et al., 2013). One of the main advantages of the use of microalgae for wastewater treatment is the diversity of removal mechanisms for different types of pollutants. Nutrients assimilation, nitrogen volatilization and phosphorous precipitation, aerobic biodegradation of organic matter, ammonium removal through nitrification, biosorption of heavy metals and pathogen disinfection due to pH fluctuations, are some of the documented mechanisms (Alcántara et al., 2015). Overall, microalgae cultivation using wastewater as the growth medium has multiple applications in biofuel production, carbon dioxide mitigation and bioremediation (Cai et al., 2013), while offering a positive impact on the production and emission of greenhouse gasses (Di Termini et al., 2011). However, one of the main drawbacks is the large area requirements to achieve higher efficiencies and removal rates than conventional systems such as activated sludge.

Numerous studies at pilot and laboratory scale level have demonstrated the potential of microalgae for the removal of different contaminants, mainly nitrogen and phosphorous (Aslan and Kapdan, 2006; González et al., 2008; Hoffmann, 1998; Park and Craggs., 2010). The main mechanisms for nutrient removal reported in the literature are via algal utilisation, nitrification/denitrification, and pH fluctuations that promote ammonia stripping or phosphorous precipitation (Cai et al., 2013). Furthermore, for nitrogen and phosphorous there is a limited range of treatment technologies that meet the standards in developed and developing countries (von Sperling and Chernicharo, 2002), being decentralized systems easier to operate with lower operational costs, making them more appropiate.

Microalgae have a high potential to be applied for the treatment of nutrient rich wastewaters due to their capacity for nutrient uptake. Consequently, microalgae systems can be used as post-treatment systems for the removal of nutrients from effluents treated in anaerobic units, which usually contain substantial amounts of nitrogen and phosphorus (Olguín, 2003). For instance, as claimed by Ruiz-Martinez et al. (2012), the use of microalgae as a post-treatment of an anaerobic membrane bioreactor can provide an excellent water quality in the effluent, while generating biogas and recovering nitrogen and phosphorous from the microalgal biomass.

1.1.2 Wastewater treatment using microalgal-bacterial systems

For microalgal systems to be competitive against other technologies such as activated sludge, the design and operation of the reactors must achieve faster and more efficient removal rates, and simultaneously strive to decreasing the area requirements, while offering lower operation costs. One of the nitrogen and carbon removal mechanisms that can achieve higher removal rates is through the symbiosis between microalgae and aerobic bacteria. This removal mechanism is achieved through the dual action of microalgae and bacteria: during the photosynthetic process microalgae assimilate CO_2 and generate oxygen, the latter can be used by heterotrophic bacteria to oxidize the organic matter and produce CO_2, while nitrifiers oxidize ammonium, creating a symbiotic relationship (Muñoz and Guieysse, 2006; Samorì et al., 2013).

This consortium between microalgae and bacteria has shown promising results for high strength and municipal wastewaters (Godos et al., 2010; González-Fernández et al., 2011a; Hernández et al., 2013; Su et al., 2012a; van der Steen et al., 2015; Wang et al., 2015; Liu et al., 2017; Maza-Márquez et al., 2017). They can achieve nitrogen removal efficiencies of between 50 and 90% without (external) aeration and phosphorous removal efficiencies of up to 60% mostly by entrapment in the biomass or chemical precipitation. Karya et al. (2013) reported 100% nitrification using synthetic wastewater in a photobioreactor (24-12 hours hydraulic retention time) using a mixed culture of algae and nitrifying bacteria without (external) aeration. The oxygen production of that setup was estimated to be 0.46 kg O_2 m^{-3}

d^{-1}, which is higher than in HRAP or stabilisation ponds (Karya et al., 2013) and the highest ammonium removal rate was 7.7 mg NH_4^+-N L^{-1} h^{-1}. Furthermore, the short hydraulic retention time makes the processes competitive with activated sludge. Vargas et al. (2016) reported 92% removal of ammonium from synthetic wastewater with an average concentration of 1214 (± 40) mg NH_4^+-N L^{-1}, being 60% removed through nitrification, while 40% was assimilated by microalgae. The advantage of photo-oxygenation compared with traditional systems is the lower operational costs.

Microalgal-bacterial consortia present themselves as a novel option for wastewater treatment. Nevertheless, there is a lack of understanding concerning the kinetic parameters and operational conditions that can enhance the removal efficiencies and rates. The use of photo-oxygenation as main source of oxygen for aerobic processes is a key aspect to increase the removal rates. However, in order to maximize these interactions, the optimum operational parameters, and their limitations and/or the negative effects of the interaction between algae and bacteria need to be determined to make these systems an attractive option for wastewater treatment.

1.2 PROBLEM STATEMENT

There is a need for municipal and industrial wastewater treatment systems that can provide high quality treated effluents, with low energy consumption and that are technically and economically easy to operate and maintain. Activated sludge systems present limitations due to their high energy requirements (mostly for aeration) and their high capital costs (Osada et al., 1991) that can account for between 45 - 75% of the total energy consumption (Ekama and Wentzel, 2008a).

On the other hand, anaerobic treatment has poor nutrient removal efficiencies and sometimes the effluent concentrations of certain parameters are even higher than in the influent (Khan et al., 2011). These effluents thus require post-treatment processes in order to remove nutrients that otherwise would promote eutrophication in the receiving surface water bodies.

Using wastewater to support algal growth in open and closed reactors, algal-bacterial interactions could help to achieve an efficient degradation. Literature has shown the potential of microalgal and bacterial consortia for the bioremediation of wastewaters with high concentrations of nutrients, especially in the agro-industrial sector and for the post-treatment of effluents from anaerobic systems without the use of external aeration (de Godos et al., 2016; González et al., 2008; Hernández et al., 2013), which reduces considerably the operational costs. The energy consumption of a HRAP using a microalgal-bacterial consortium treating a primary settled effluent is approximately 0.023 kWh per m^3 of wastewater treated (Alcántara et al., 2015), while the same effluent would require 0.33-0.60 kWh m^{-3} (Plappally and Lienhard, 2012) in an activated sludge system. However, the area requirements of HRAP systems are larger than for conventional activated sludge and anaerobic treatment, due to the longer required HRT and shallow depths. Therefore, the investment costs for HRAP in relation with the area increases the total cost of the system. Microalgal-bacterial systems should thus aim to reduce the HRT while maintaining the removal efficiency in order to be competitive with conventional technologies. As a result, it is proposed to develop and evaluate a novel bio-treatment system for open ponds, like high-rate algae ponds (HRAPs), that can make use of consortia composed of microalgae and bacteria (the so called Photo-Activated Sludge (PAS)) for the treatment of nitrogen from wastewater.

The co-cultivation of the key groups of microorganisms can allow a faster degradation of contaminants and the enhancement of physiological mechanisms that a single species or strain cannot easily carry out (Brenner et al., 2008). However, there is a lack of understanding concerning the interactions that govern the carbon and nitrogen removal mechanisms of the symbiosis, and even regarding the effects of key parameters such as the kinetics of the microorganisms involved (heterotrophs, autotrophs and chemoautotrophs). In order to maximize the nutrient removal efficiencies, it is necessary to determine which kinetics and stoichiometry parameters are more sensitive for the consortia of algae and bacteria. For instance, it is important to determine the role of the maximum growth rate of each of the different microorganisms (μ_m) (nitrifiers, heterotrophic bacteria and algae), and the different operational conditions that have the higher impact on the nutrients removal.

The identification of these parameters can lead to the determination of the (optimal) design and operational parameters (mostly HRT and SRT) that affect the consortia and contribute to the development of design principles for microalgal-bacterial reactors. For instance, the development of an open reactor that can achieve an optimum operation using microalgal-bacterial consortia implies the achievement of a sustainable photo-oxygenation for the oxidation of ammonium (nitrification) and organic matter. Moreover, the understanding of the symbiosis could perform the double task of pollutant degradation, as well as the commercial production of by-products, contributing to the mitigation of CO_2 emissions (Subashchandrabose et al., 2011) and the reduction of the footprint and land requirements.

The main objective of the present research is to maximize the efficiency of microalgal-bacterial consortia for the removal of nitrogen from wastewater rich in ammonium nitrogen, e.g. effluents from anaerobic digestors. The understanding of the symbiosis between microalgae and bacteria and the effect of the key design and operational parameters (SRT and HRT) affecting their interactions will help to define the optimal conditions for open photobioreactors such as HRAPs.

The research will focus on investigating the kinetics of the microorganisms involved (heterotrophs, autotrophs and phototrophs) in order to optimize the removal efficiency of the microalgal-bacterial consortia. The phototrophic organisms that this thesis will focus on are eukaryotic algae and prokaryotic cyanobacteria. The overall work will contribute to the development of design criteria for high rate algae ponds using the PAS system for wastewater treatment as a simple, yet innovative, technology with low energy requirements, high removal efficiencies of organic compounds and nutrients.

1.3 Outline of thesis

This thesis has been structured in 8 chapters (Figure 1.1). **Chapter 1** is a general introduction on the background of the research, highlighting the benefits of natural systems for wastewater treatment, focusing on the implementation of microalgal-bacterial consortia in standard algal systems such as HRAP and the research needs for the development of microalgal-bacterial systems.

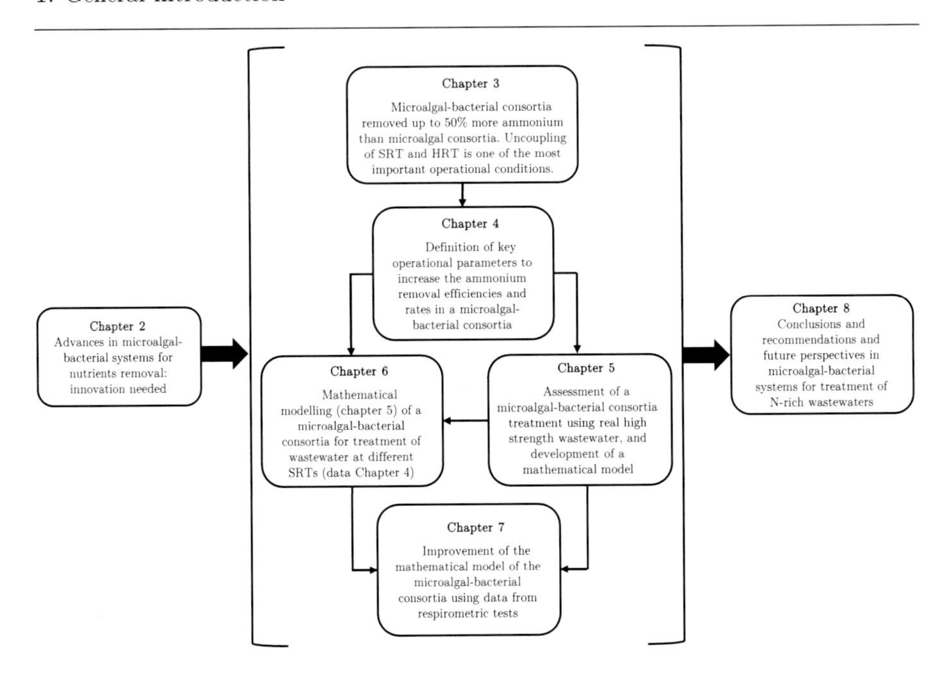

Figure 1.1. Thesis structure and connection among the different chapters

Chapter 2 presents the state of the art of the microalgal-bacterial interactions within wastewater, nutrient removal, photobioreactor configurations, and growth limitations.

In **Chapter 3** the benefits of using microalgal-bacterial consortia over solely algal consortia is demonstrated in a flat panel sequencing photobioreactor, and the differences in the ammonium removal rates and pathways at different loading rates and HRT are studied. **Chapter 4** studies the effect of the SRT on the removal mechanisms of the microalgal-bacterial consortia and determines the relationship between the removal rates and the length of the SRT in a sequencing batch photobioreactor. **Chapter 5** demonstrates the feasibility of the microalgal-bacterial consortia to treat high ammonium strength wastewater in a sequencing batch photobioreactor. Furthermore, the data was used to propose and calibrate a mathematical model to describe the microalgae and bacteria processes for the removal of ammonium and microalgal-bacterial growth.

Chapter 6 presents the result of the calibration and validation of the mathematical model proposed in Chapter 5 at different SRTs, and using the laboratory data of Chapter 4. **Chapter 7** demostrates the ability of algae to store nitrogen intracellularly. The mathematical model developed in Chapter 5 and 6 was updated including the processes related with nitrogen storage and phototrophic growth on nitrogen storage compounds.

Chapter 8 summarizes the results of this research, and discusses the future perspectives for the proposed technology. It also gives recommendations for practical applications and future research.

2

MICROALGAL-BACTERIAL CONSORTIA FOR WASTEWATER TREATMENT: A REVIEW

2.1 MICROALGAL-BACTERIAL CONSORTIA

The basis of the Photo-Activated Sludge system (PAS) for the treatment of nitrogen in wastewater is the consortium between microalgae and bacteria (Figure 2.1). Microalgae and bacteria co-habit in freshwater, wastewater and marine systems. Symbiosis among aerobic bacteria and microalgae for treatment of wastewater was first reported by Oswald et al. (1953) in oxidation ponds. One of the interactions reported is the exchange of oxygen: the oxygen produced by the microalgae, through photosynthesis, is used by aerobic bacteria (heterotrophic and nitrifiers) to oxidize organic matter and ammonium. Heterotrophic bacteria produce carbon dioxide through respiration and oxidation of organic matter, which can be taken up as a carbon source by the microalgae. In the case of nitrogen, after nitrate is produced, it can be taken up by microalgae as a source of nitrogen, or further denitrified by bacteria when anoxic conditions are met, usually during dark periods, or dark zones within the reactor. These interactions create a synergistic relationship between microalgae, heterotrophs and nitrifiers in which the required oxygen is supplied by microalgae. The aeration supplied by microalgae is defined as photosynthetic oxygenation. The term was first defined by Oswald et al. (1953) as "*production of oxygen through the action of light on the chloroplastic tissue of microscopic green plants, growing dispersed in the aqueous medium*".

The symbiosis has been reported to occur in waste stabilization ponds, oxidation ponds and high rate algae ponds (HRAP). Zhou et al. (2006) reported removal of nutrients through nitrification/denitrification in high rate algae ponds treating rural domestic wastewater. About 50% of the nitrogen was removed through nitrification/denitrification, followed by algae assimilation and sedimentation. In the case of phosphorus, the main removal mechanisms were through algae assimilation followed by chemical precipitation.

Additional to the removal of nutrients, a consortium of algae and bacteria is able to remove hazardous pollutants, as reviewed by Muñoz and Guieysse (2006). Pollutants such as acetonitrile were found to be removed at a rate of 2300 mg $L^{-1}d^{-1}$ by a consortium of *Chlorella sorokiniana* and a bacterial consortium suspended in a stirred tank reactor. Safonova et al. (2004) reported the removal of different xenobiotic compounds through a consortium of algae and bacteria. They observed different removal efficiencies for phenols (85%), anionic surfactants such as secondary alkane sulfonates (73%), oil spills (96%), copper (62%), nickel (62%), zinc (90%), manganese (70%) and iron (64%). The consortia used consisted of the algal strains *Chlorella* sp., *Scenedesmus obliquus*, *Stichococcus* and *Phormidium* sp. and of bacterial strains such as *Rhodococcus* sp., *Kibdelosporangium aridium* and two other unidentified bacterial strains. The removal mechanisms were the association between the oil degrading bacteria and the algal strains, the ability of algae to supply oxygen and at the same time the ability of aerobic bacteria to degrade hydrocarbons.

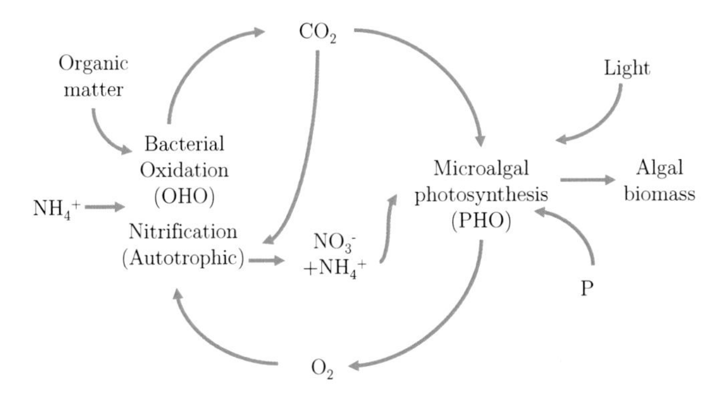

Figure 2.1. Microalgae and bacterial oxidation interactions in a microalgal – bacterial consortia. Source: Adapted from Muñoz and Guieysse (2006). OHO: Heterotrophic organisms, PHO: phototrophic organisms, and P: phosphorous.

2.1.1 Interactions within microalgal-bacterial consortia

The interactions between algae and bacteria are not limited to the exchange of carbon dioxide and oxygen. On the opposite, the interactions can be either mutualism, parasitism or commensalism (Ramanan et al., 2016). As a result, algae and bacteria are able to change their physiology and metabolism (Ramanan et al., 2016).

There are several studies showing the benefits and negative effects of bacteria and algae when present in consortia (Unnithan et al., 2014). Algae can either promote bacterial growth through the release of organic exudates (Abed et al., 2007), nutrient exchange as result of algal lysis (Unnithan et al., 2014), or decreased algal growth through the release of algicidal substances by bacteria (Fukami et al., 1997) and/or pH fluctuations as a result of the photosynthesis. Kirkwood et al. (2006) reported how the production of exudates by cyanobacteria did not completely inhibit bacterial growth, but instead were used as substrate in a consortium of heterotrophic bacteria and cyanobacteria treating pulp and paper wastewater. In addition, the study revealed that the exudates also enhanced the removal of dichloroacetate, and at the same time affected the removal of phenolic compounds.

Choi et al. (2010) reported the negative effect of cyanobacteria on the nitrification rates in a bioreactor growing only nitrifiers. The presence of algae and cyanobacteria in the autotrophic bioreactor inhibited the maximum nitrification by a factor of 4, however, the ammonium was still efficiently removed (Choi et al., 2010). Other negative effects of microalgae on bacteria are the increase in pH due to the photosynthetic activity and high dissolved oxygen concentration. The fast growth rate of microalgae can create a high density in the culture that led to the increase of dark zones, in which microalgae can perform respiration and diminish the amount of oxygen for bacteria (Muñoz and Guieysse, 2006).

On the opposite, there are also microalgae growth-promoting bacteria (MGPB). As the name states, these bacteria enhance the growth of microalgae. De Bashan et al. (2004) demonstrated how the bacterium *A. brasilense* boosted the growth of *Chlorella sorokiniana*, which lead to an effluent with less nitrogen and phosphorus. Additionally, the consumption of oxygen by the aerobic bacteria helps to prevent oxygen saturation conditions.

The presence of bacteria in microalgal cultures improves the flocculation of suspended algae. Some studies have reported the improvement in the settling characteristics of the biomass in microalgal-bacterial cultures through the formation of granules or aggregates (Gutzeit et al., 2005; Lee et al., 2013; Van Den Hende et al., 2014). The formation of flocs in an algal-bacterial consortium is promoted by the bacterial exopolymers, increasing the aggregation and stabilizing the already existing aggregates, while increasing settleability (Subashchandrabose et al., 2011). Algal-bacterial flocs vary from 50 µm to 1 mm, but the predominant size is between 400 - 800 µm (Gutzeit et al., 2005). Tiron et al. (2017) reported the development of granules or as the author calls them "activated algae flocs", for this already formed algal flocs and the bacterial population already present in the raw dairy wastewater were used as inoculum. The developed activated algae granules had a size between 600 – 2000 µm, and a settling velocity of 21.6 (± 0.9) m h^{-1} (Tiron et al., 2017). Figure 2.2 presents an example of an activated algae granule. This positive effect tackles one of the drawbacks of solely algal systems: efficient biomass harvesting. Tiron et al. (2017) the formation of the granules was achieved in a 1.5 L photobioreactor operated as sequencing batch using diluted pretreated dairy wastewater (15.3 – 21.8 mg NH_4^+-N L^{-1}) with an HRT between 96 - 24 hours.

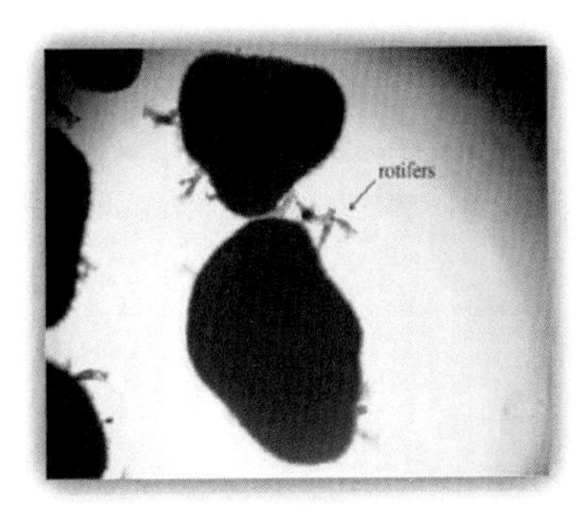

Figure 2.2. Algae granules containing the algae strains: Chlorella sp. and Phormidium sp. (Tiron et al. 2017)

Despite some of the negative interactions, the consortium between microalgae and bacteria enhances the removal of nutrients and other pollutants. The synergistic relationship provides sturdiness to overcome extreme environmental conditions and fluctuations due to operational changes. The complexity of these interactions needs to be understood in order to maximize the positive effects to develop culture conditions that enhance wastewater treatment.

2.1.2 Nutrient removal by microalgal-bacterial consortia

The main difference between an algal system and a microalgal-bacterial consortium in terms of nitrogen removal is the removal pathways. In algal systems, assimilation into the biomass and ammonium volatilization due to pH fluctuations are the two main removal mechanisms. In microalgal-bacterial consortia these are not the only removal mechanisms, but another important pathway of nitrogen removal is nitrification, as nitrifiers can make use of the oxygen produced by the microalgae. The exchange of oxygen and carbon dioxide allows the efficient removal of organic matter and nitrogen by heterotrophic and nitrifying bacteria. Furthermore, open and closed photobioreactors contain dark zones in which anoxic conditions allow denitrification by anoxic heterotrophic (denitrifying) bacteria.

Phosphorus can be removed from the water either by chemical or microbiological mechanisms. Like nitrogen, phosphorus is an essential nutrient for microalgae. Phosphorus is taken up by algae preferably in the forms of $H_2PO_4^-$ and HPO_4^{2-}, and incorporated into the cell through phosphorylation (transformation into high energy organic compounds) (Martínez et al., 1999). However, there is no a clear description in the literature about how the phosphorous removal is achieved in waste stabilization ponds, as the reasons are not well understood (Powell et al., 2008). The chemical mechanism of phosphorus removal is through precipitation. This mechanism depends on the pH and the dissolved oxygen concentration in the bulk liquid. At high pH and dissolved oxygen concentrations, phosphorus will precipitate (Cai et al., 2013). de Bashan and Bashan (2004) presented a review of the different forms of phosphorus precipitation. Usually it can occur at pH higher than 9, depending on the concentrations of the different ions and P. Due to the fact that phosphorus does not exist in gaseous form (like atmospheric nitrogen which eventually could be fixed by algae) and that it can be easily bound with other ions, it is the most important growth limiting factor in microalgae cultivation, besides light (Grobbelaar, 2008). Phosphorus assimilation is the main biological mechanism of removal in algal systems. Di Termini et al. (2011) achieved phosphorus removal between 80 - 90% in outdoor and indoor closed photobioreactors through microalgae assimilation.

Several authors have reported the use of microalgal-bacterial consortia for nutrient (nitrogen and phosphorous) removal from real or synthetic wastewater using different types of photobioreactors (Subashchandrabose et al., 2011). The different studies showed nitrogen removal efficiencies were between 100% and 15%, whereas the phosphorous removal efficiencies were between 90% and 31.5% (Subashchandrabose et al., 2011).

The symbiosis between microalgae and bacteria offers a large potential for the treatment of nutrient rich wastewaters, although some aspects need to be taken into account, as they determine the nutrient removal efficiencies or the nutrient removal pathways. The selection of a particular strain for wastewater treatment is a decisive step when engineering a consortium of microalgae and bacteria. In open ponds, there is a natural selection of the microalgae species, which depends on the organic load of the wastewater, species interactions, seasonal environmental conditions, competition and interactions among the microorganisms present in the culture (Riaño et al., 2012). Natural selection of microalgae within a microalgal-bacterial consortium allows to achieve higher efficiencies as there are no inhibitory effects by the source of the wastewater.

González-Fernández et al. (2011a) compared the removal efficiency of 4 ponds using microalgal-bacterial consortia for the treatment of pig slurry. The ponds differed in terms of operational conditions (optimal and real conditions), and source of the slurry (anaerobically digested or fresh). The three reported removal mechanisms were nitrification/denitrification, stripping and biomass uptake. Among these three, the main driving force of removal depended on the substrate source. The NH_4^+-N/COD ratio of the substrates was responsible for the different removal rates and the main removal pathway. The anaerobic digested slurry had a ratio of 0.46 NH_4^+-N/COD, whereas the fresh slurry had a NH_4^+-N/COD ratio of 0.13. Since the organic matter in the anaerobically digested slurry is more recalcitrant, the oxygen is more likely taken up for nitrification, reason why nitrification rates were higher for ponds fed with anaerobically digested slurry (González-Fernández et al. 2011a).

Molinuevo-Salces et al. (2010) compared open and closed configurations and the results showed that even though ammonium was completely removed, the removal mechanisms were different. In the open configuration the biomass uptake was between 38 - 47%, while 52 - 29% was nitrified/denitrified. In the closed reactor 10.5% was volatilized and 11.3% nitrified, 41% nitrified/denitrified and 31.3% taken up by algae (Molinuevo-Salces et al., 2010). About 80% of the phosphorous was removed regardless the configuration.

Ammonium removal through nitrification/denitrification as main removal mechanism in microalgal-bacterial systems has the advantage of achieving faster removal rates in comparison with solely algal systems, especially for high concentrated effluents from industrial sectors. Wang et al. (2015) used microalgal-bacterial consortia to treat anaerobically digested swine manure with ammonium concentrations up to 297 (± 29) mg NH_4^+-N L^{-1} (value after 3 times dilution) in a sequencing batch photobioreactor (4 days hydraulic retention time), achieving a 90% total nitrogen (TN) removal efficiency, from which 80% was removed through nitritation/denitritation without any external aeration. Furthermore, Manser et al. (2016) reported the successful combination of microalgae, ammonium-oxidising bacteria (AOB) and anammox in a sequencing batch photobioreactor achieving ammonium oxidation to nitrite at a rate of 7.0 mg NH_4^+-N L^{-1} h^{-1} in the light periods, and during the night periods in which anoxic conditions were achieved, about 82% of the nitrite was reduced by anammox bacteria.

Table 2.1 Nutrient removal using a microalgal-bacterial consortia for different types of wastewater and using different types of reactors. Source: Subashchandrabose et al. (2011).

Cyanobacterium/ microalga	***Bacterium***	***Source of waste water***	***Nutrients and removal efficiency***	***System - reactor used***
Spirulina platensis	Sulfate-reducing bacteria	Tannery effluent	Sulfate 80% (2000 mg/L)	High rate algal pond (HRAP)
Chlorella vulgaris	*Azospirillum brasilense*	Synthetic wastewater	Ammonia 91% (21 mg/L) Phosphorous 75% (15 mg/L)	Chemostat
Chlorella vulgaris	Wastewater bacteria	Pretreated sewage	DOC 93% (230 mg C/L Nitrogen 15% (78.5 mg/L)	Photobioreactor pilot-scale

Cyanobacterium/ microalga	*Bacterium*	*Source of waste water*	*Nutrients and removal efficiency*	*System - reactor used*
			Phosphorous 47% (10.8 mg/L)	
Chlorella vulgaris	*Alcaligenes sp.*	Coke factory wastewater	NH_4^+ 45% (500 mg/L) Phenol 100% (325 mg/L)	Continuous photobioreactor with sludge recirculation
Chlorella vulgaris	*A. brasilense*	Synthetic wastewater	Phosphorous 31.5% (50 mg/L) Nitrogen 22% (50 mg/L)	Inverted conical glass bioreactor
Chlorella sorokiniana	Mixed bacterial Culture from an activated sludge process	Synthetic wastewater	Phosphorous 86% (15 mg/L) Nitrogen 99% (180 mg/L)	Tubular biofilm photobioreactor
Chlorella sorokiniana	Activated sludge bacteria	Pretreated piggery wastewater	TOC 86% (645 mg/L) Nitrogen 87% (373 mg/L)	Glass bottle
Chlorella sorokiniana	Activated sludge bacteria	Pretreated swine slurry	TOC 9-61% (1247 mg/L) Nitrogen 94-100% (656 mg/L) Phosphorous 70-90% (117 mg/L)	Tubular biofilm photobioreactor

Cyanobacterium/ microalga	*Bacterium*	*Source of waste water*	*Nutrients and removal efficiency*	*System - reactor used*
Chlorella sorokiniana	Activated sludge bacteria	Piggery wastewater	TOC 47% (550 mg/L) Phosphorous 54% (19.4 mg/L) NH_4^+ 21% (350 mg/L)	Jacketed glass tank photobioreact or
Euglena viridis	Activated sludge bacteria	Piggery wastewater	TOC 51% (450 mg/L) Phosphorous 53% (19.4 mg/L) NH_4^+ 34% (320 mg/L)	Jacketed glass tank photobioreact or
Microalgae present in tertiary stabilization pond treating domestic wastewater	Bacteria present in tertiary stabilizatio n pond treating domestic wastewater	Piggery wastewater	COD 58.7% (526 mg/L) Total Kjeldahl Nitrogen 78% (59 mg/L)	High rate algal pond (HRAP)

2.1.3 Microalgal-bacterial systems and configurations

Algal wastewater treatment systems can be divided in open and closed photobioreactors. According to the reactor geometry, closed photobioreactors can be divided into: (i) vertical columns, (ii) tubular reactors and (iii) flat panel reactors (Wang et al. 2012). Open reactors can be listed into: (i) waste stabilization ponds (WSP), (ii) raceway ponds and (iii) high rate algae ponds (HRAP). Figure 2.3 presents a scheme of the three most used photobioreactors for algal cultvations.

Currently, open systems are the most used type for wastewater treatment and biomass cultivation using microalgae (Carvalho et al., 2006; Wang et al., 2012) due to their low investment and maintenance cost and easiness to scale up (Cai et al., 2013). Closed systems are mostly used for sensitive microalgae strains, products vulnerable to microbial degradation or when the harvested biomass is aimed at direct human consumption such as for cosmetics or nutritional supplements (Carvalho et al., 2006). Closed systems have a higher light harvesting, thus biomass production can achieve a higher population density, however the investment and maintenance costs are higher compared with open systems (Carvalho et al., 2006).

HRAP are the most efficient open systems as they are operated with a higher depth in comparison with the other options. HRAP are raceway type ponds with depths between 0.2 - 1 m. They can treat up to 35 g BOD m^{-2} d^{-1} compared with 5 - 10 BOD m^{-2} d^{-1} in waste stabilisation ponds (Muñoz and Guieysse, 2006). However, light penetration in such reactors is limited by the depth or solids concentration. Furthermore, open and closed systems both require large areas for operation in order to either efficiently remove the contaminants or to achieve high biomass production. Therefore, the reactor selection and the growth medium composition depends on the objective of the system.

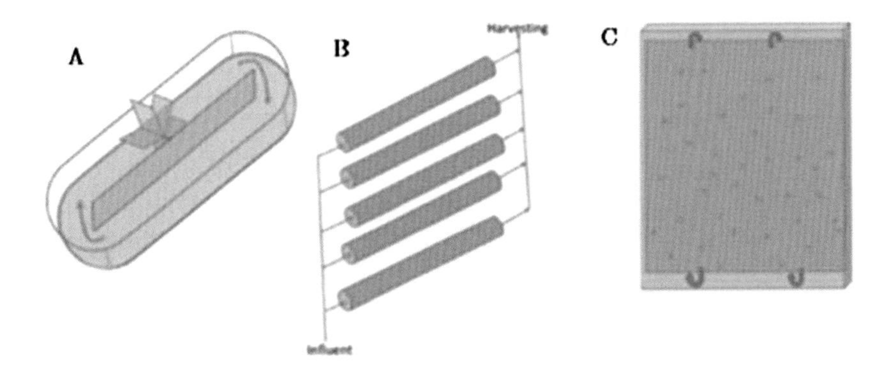

Figure 2.3. The three most used algal system configurations. A) High rate algae pond, B) Closed tubular photobioreactor, and C) Flat panel airlift reactor Source: (Wang et al., 2018).

2.1.4 Limiting and operational conditions of microalgal-bacterial photobioreactors

There are several factors that can affect the growth of algae and bacteria, especially when using wastewater as growth medium, since there are many substances, compounds and factors to take into account. In open and closed photobioreactors there are physical, chemical, biological and operational factors that can limit the growth of microalgae (Borowitzka, 1998). Among those, the parameters that have a strong effect on the efficiency of microalgae and bacteria when treating wastewater are: pH, light intensity, temperature, dissolved carbon dioxide, nutrients, mixing, dilution and algae harvesting (Borowitzka, 1998; Rawat et al., 2011).

In terms of operation, different operational parameters have an effect on the cultivation of microalgae and bacteria separately. Therefore, special attention should be given when combining these two groups of microorganisms. One of the most critical operational parameters is the biomass retention time, which in the case of a consortium can be determined by the influent flow rate, and whether there is biomass recirculation. Solid retention time (SRT) and hydraulic retention time (HRT) influence the biomass concentration and the overall productivity of the microalgal-bacterial systems (Valigore et al., 2012). This PhD research focused on open photobioreactors such as high rate algae ponds. For this reason, the implications of some of the factors limiting algal and bacterial growth in high rate open algal ponds are described below.

Light

Light is the energy source to perform photosynthesis, allowing microalgae growth. Hence, the uptake efficiency of light is crucial for the productivity of algal biomass and photo-oxygenation. Microalgae can absorb only a fraction of the irradiance, between 400 - 700 nm. This range is called the photosynthetically-active radiation (PAR). Open ponds obtain this irradiance from the sun, hence the ponds are shallow in order to allow a maximal light penetration. Height it is not the only limitation for the light irradiance, attenuation by the biomass itself is another factor, which can increase when co-cultured with bacteria, and the fact that light can be easily absorbed by other materials or substances (Fernández et al., 2013; Jeon et al., 2005). Dense and concentrated cultures present mutual shading, reducing the light intensity from the illuminated surface to the centre of the reactors, which increase the dark zones and consequently microalgal respiration (Chen et al., 2011; Fernández et al. 2013). Due to this, microalgae are exposed to light/dark zones. For instance, in open ponds except for the upmost thin layer, the irradiance in the pond is below the photo-compensation point for algal growth (Barbosa et al., 2003), as a result of this photosynthetic rates decrease, as well as algal growth. This effect can be compensated by a good mixing which allows the cells to be exposed to a sufficient amount of irradiance (Chen et al., 2011). In open ponds, usually the mixing is provided by a paddle wheel, while aeration is usually applied in closed photobioreactors.

Indoor cultures and closed photobioreactors use other sources of light different from sunlight. For instance, high pressure sodium lamps, tungsten-halogen lamps, fluorescent tubes and light emitting-diodes (LED lights). Although, these lamps provide a reliable source of energy, the disadvantages are the high power consumption and high operational costs, and they do not contain the full spectrum of light energy (Chen et al., 2011). On the other hand, sunlight is free and holds the full spectrum of light energy.

pH

pH is one of the most important parameters in microalgal cultures, as it determines the solubility of carbon dioxide, removal of other nutrients like P and N, and most importantly it affects the metabolism of the microalgae (Becker, 1994). Furthermore, pH fluctuations can inhibit bacterial activity such as autotrophic and heterotrophic bacteria. Fluctuations of pH in microalgae cultures are a consequence of the processes of photosynthesis and respiration during the light and dark periods, respectively. During the day, the pH increases due to the assimilation of CO_2 and the release of OH^-. pH values of up to 10 have been reported after the depletion of NO_3^- and CO_2 (Becker, 1994). Increments of the pH are limited in some cases by the respiration of the different microorganisms. Additionally, nitrogen removal through nitrification has an effect on the pH fluctuations, since the pH decreases during this process due to the release of H^+. Therefore, the addition of ammonium can help to reduce the pH increment (Larsdotter, 2006), making it a good option for pH control in open ponds. Also, the addition of CO_2 can help to control the pH as shown by Park and Craggs (2010).

pH values can affect the growth of microalgae and therefore the removal of nutrients, this can vary for the different strains. Some algae such as *Microcystis aeruginosa* and *Anabena spiroides* have growth limitations and inhibition when exposed to a pH below 6 (Wang et al., 2011). pH fluctuations can also determine the removal of N and P, as higher pH causes ammonium volatilization and phosphorus precipitation. When this occurs faster than the uptake by algae, it leads to algal growth limitation due to the lack of nutrients. Therefore, pH control strategies must be developed in order to avoid possible negative effects caused by drastic pH fluctuations.

In the case of nitrifiers, the growth is suppressed when the pH is not within the 7 to 8 range (Ekama and Wentzel, 2008a). Nitrification performed by aerobic bacteria release hydrogen ions, reducing the alkalinity of the bulk liquid. Stoichiometrically, for every 1 mg free and saline ammonia (FSA) nitrified, 7.14 mg alkalinity ($CaCO_3$) is consumed (Ekama and Wentzel, 2008a). When alkalinity is lower than 40 mg L^{-1} in activated sludge systems, the pH decreases to low values, compromising the nitrification rates and settleability characteristics of the sludge (Ekama and Wentzel, 2008a). In systems working with algae and bacteria, the pH drop by nitrification can be counterbalanced by photosynthetic activity. Also denitrification recovers alkalinity, which occurs under anoxic conditions. In algal-bacterial systems, dark conditions guarantee the absence of oxygen production by algae, instead algae respire releasing CO_2, which helps to decrease the pH. Based on this, it is evident that the balance in terms of alkalinity between microalgae and bacteria is important.

Hydraulic retention time

Hydraulic retention time controls the nutrient loading rates, which at the same time will control the productivity and nutrient removal rate of an algae system. In an open pond with well mixed and steady-state conditions, the productivity is governed by the dilution rate and the depth of the pond. The HRT corresponds to the reciprocal of the dilution rate. In algal ponds and HRAP, the HRT is the same as the solids retention time (SRT), since it is not common to recirculate the biomass, as the harvesting of algal biomass is one of the biggest challenges due to their low cell size (Lee et al., 2013). Therefore, in order to achieve complete removal rates of pollutants, it is common practice to operate algal systems at a HRT between 2-8 days and depths between 0.2 -0.5 m (Shilton, 2006). Due to seasonal variations, it is recommended to vary the HRT, as the temperature changes limit or enhance the growth rates.

Furthermore, shorter HRT in algal systems enhance the biomass production (Oswald et al., 1953; Takabe et al., 2016). Valigore et al. (2012) compared different HRT (from 8 -1.4 days) in a microalgal-bacterial culture, concluding that a shorter HRT enhanced the biomass productivity. However, a shorter HRT can decrease the nutrient removal rates in microalgal-bacterial systems, especially when it can promote wash out of the biomass. An optimum HRT enhances nutrient removal by allowing the proper growth of algal-bacterial populations, which will promote faster nitrification rates, especially since the growth rate of nitrifying microorganisms is low, i.e. μ_m=0.45 d^{-1} at 20ºC (Ekama and Wentzel, 2008a). Therefore, the HRT must be chosen depending on the objective, whether the maximization of the biomass production or the treatment of wastewater. Also, it must be taken into account that due to the depth of the HRAP, a longer HRT will result in larger areas, therefore optimization of this parameter is crucial for algal systems.

Solids retention time

When working with a consortium of microalgae and activated sludge bacteria for nutrient and organic matter removal through photo-oxygenation, the sludge retention time plays an important role within the operational parameters. In fact, it is the most fundamental and important decision for the design of activated sludge systems (Ekama and Wentzel, 2008b). Sludge retention time controls the growth of the microorganisms, and corresponds to the relation between the volume of the reactor and the waste biomass flow from the reactor. Therefore, the sludge production in activated sludge systems decreases with the increase of the SRT (Ekama and Wentzel, 2008b). On the other hand, for suspended algae systems, the algae biomass production is controlled by the HRT. This parameter controls the biomass concentrations, which will affect the light utilization by microalgae (Lambeert-Beer law).

Figure 2.4 presents the productivity curve for a flat panel reactor for different biomass concentrations and light intensity. The optimal concentration ($C_{x,opt}$), where the biomass production is at the maximum, will depend on the efficient use of light. This is achieved when the light at the back of the reactor equals the compensation point for microalgae growth. For lower concentrations, the light will pass through the reactor un-used, whereas for higher values, the light will not be able to reach the bottom/back of the photobioreactor (Janssen and Lamers, 2013). Therefore, there is a need for optimum SRT and HRT combinations to achieve a microalgal-bacterial biomass concentration that allows complete nitrification by ensuring sufficient oxygen without biomass wash-out.

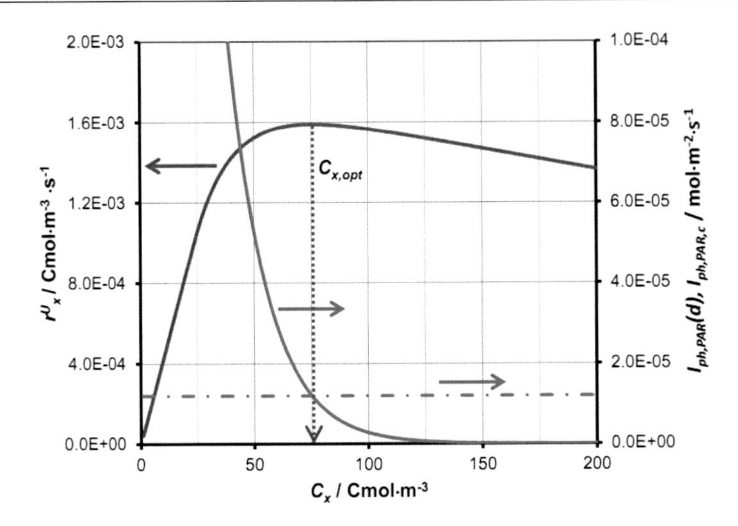

Figure 2.4. Volumetric productivity of a photobioreactor r^{u}_{x} as a function of the biomass concentration C_x. Light intensity at the back of the reactor $I_{ph,PAR}$ (d) and the compensation light intensity $I_{ph,\ PARc}$, are also shown. Source: Janssen and Lamers (2013)

Valigore et al. (2012) concluded that biomass recycling at a SRT higher than the HRT reduces the wash-out of the microorganisms present in the reactor. Therefore, an appropriate SRT will ensure the successful growth of nitrifiers (slower growing microorganisms in activated sludge) and in addition guarantees light availability for photo-oxygenation. The recommended ranges of SRT values for complete nitrification are divided in two: (i) intermediate, between 10 to 15 days, this range ensures complete nitrification, and (ii) long sludge age refers to more than 20 days, for which the production of sludge is low with a rather inactive sludge (Ekama and Wentzel, 2008b).

The sludge retention time also plays a role in the floc formation, since longer SRT and biomass recirculation enhances the biomass settleability and floc formation (Gutzeit et al., 2005; Medina and Neis, 2007; Valigore et al., 2012). It was reported that settleability of algal-bacterial biomass increased from 13 to 93% when the SRT increases up to 40 days (Valigore et al., 2012). Additionally, Gutzeit et al. (2005) achieved during a period of 18 months a flocculent algal - bacterial biomass with excellent sedimentation characteristics, using a SRT between 20 - 25 days. On the other hand, longer SRT promote algal death due to high solids concentrations, which limits the light penetration and creates higher dark zones increasing the respiration activity (Oswald et al., 1953). Since HRT and SRT can operationally define the removal rate, biomass characteristics and productivity, it is essential to further investigate different conditions of these two in order to define the operational conditions for novel algal-bacterial based wastewater treatment systems.

2.2 Microalgal-bacterial modelling

Modelling of processes in wastewater treatment has the advantage of getting insight into the performance of the technology, evaluation of possible scenarios for upgrading, evaluation of new plant design, support to the decision making related with operational conditions and personal training (van Loosdrecht et al., 2008). Modelling of microalgae systems, more specifically for open ponds, has to take into account several factors, such as light, wind, stripping of ammonia and carbon dioxide, as well as biological and hydrodynamic processes (Gehring et al., 2010). There are several models which focus on different microalgae processes, for instance on the net growth of microalgae (Decostere et al., 2013; Solimeno et al., 2015; Wágner et al., 2016), models dealing with light limitation and photosynthesis rates (Yun and Park, 2003), kinetics of nutrient removal (Kapdan and Aslan, 2008), pigments dynamics and respiration (Bernard, 2011) and dissolved oxygen rates (Kayombo et al., 2000).

In the case of activated sludge, bacteria are mostly modelled by a set of models (ASM1, ASM2 and ASM3, ASM3, ASM2d, ASM3-bio-P) developed by task groups of the International Water Association (IWA) and the metabolic model developed at Delft University of Technology (Gernaey et al., 2004). The activated sludge model No. 1 (ASM1) (Henze, 2000) is considered the reference model. It describes the removal of organic carbon compounds and nitrogen, while consuming oxygen and nitrate as electron acceptors. Additionally, it describes the sludge production and has adopted the chemical oxygen demand (COD) as measurement unit for organic matter (Gernaey et al., 2004). Furthermore, similar to ASM1, ASM3 was developed to correct the deficiencies of the ASM1 model. The main difference of the ASM3 model is the inclusion of the intracellular storage process of readily biodegradable COD, for the slower conversion from readily biodegradable into slowly biodegradable organic matter (Gernaey et al., 2004; van Loosdrecht et al., 2008). Other models include biological phosphorus removal, i.e ASM2d and the TUDelft model (van Loosdrecht et al., 2008).

As mentioned in previous sections, usually in open ponds that are treating wastewater, not only microalgae play a role in the removal of nutrients and biomass production, but at the same time, heterotrophic and nitrifying bacteria carry out different processes like oxidation of organic matter, nitrification, denitrification and respiration (Figure 2.1). Therefore, they make the system more complex as those microorganisms and their associated parameters and variables should be taken into account. Furthermore, models describing these complex relationships should be based on the microalgae models and activated sludge models. Models describing the relationships of algal-bacterial consortia in open ponds have been reported at first by Buhr and Miller (1983). Their objective was to develop a mathematical model for high rate algal-bacterial wastewater treatment systems. This model takes into account the algal and bacterial growth, light limitation, and solution equilibrium related with the pH and mass balances. The variations of pH, DO and substrate concentrations along the pond length were evaluated under different feed loads and hydraulic residence times. Later on, Gehring et al. (2010) developed a model to simulate the processes in a waste stabilisation pond. The activated sludge model No. 3 (ASM3) was used as a basis. The new components were the integration of algae biomass and gas transfer processes for oxygen, carbon dioxide and ammonia depending on wind velocity. Furthermore, it had the possibility to model the algae concentrations based on measured Chlorophyll-*a*, light intensity and total suspended solids (TSS) measurements (Gehring et al., 2010). However, modelling of nitrification and denitrification was not considered in the simulations performed by Gehring et al. (2010) because the experimental data did not show any nitrification or denitrification rates. Therefore, the model was not evaluated under the two conditions of nitrification and algal growth.

There are in the literature some models focused on algal-bacterial consortia (Solimeno et al., 2017; van der Steen et al., 2015; Wolf et al., 2007; Zambrano et al. 2016). Solimeno et al. (2017) developed the BIO-ALGAE model for suspended microalgal-bacterial biomass, which was an updated version of the algal model proposed by the same author (Solimeno et al., 2015). The model was calibrated and validated, reporting good results on the prediction of biomass characterization. Furthermore, it identified the light factor as one of the most sensitive parameters for microalgal growth. The model takes into account the algal growth on carbon and nutrients, gas transfer to the atmosphere, photorespiration and photoinhibition.

The PHOBIA model was developed by Wolf et al. (2007) at the Delft University of Technology for microalgal-bacterial biofilms. It includes the modelling of different kinetic mechanisms of phototrophic microorganisms, such as internal polyglucose storage, growth in darkness, photoadaptation and photoinhibition, as well as nitrogen preference (Wolf et al., 2007). These models can serve as a basis for the development of further models whose aim is to explain and describe the microalgae-bacteria symbiosis for their cultivation for wastewater treatment in suspended cultures. For this reason, there is still a need for models calibrated and validated with longer data sets or at different operational conditions treating diverse types of wastewaters.

2.3 Aims of this PhD research

The aim of this research is to maximize the efficiency of microalgal-bacterial consortia for nitrogen removal. This is intended through the understanding of the symbiosis between microalgae and bacteria and of the operational parameters SRT and HRT, which have a great effect on the consortia.

The objective is to define the optimal conditions for an innovative treatment called Photo-Activated Sludge (PAS) system. For this, the PhD focuses on the investigation of the kinetic parameters of the microorganisms involved (heterotrophs, autotrophs and photoautotrophs) in a microalgal-bacterial consortium for treatment of anaerobic effluent, in order to optimize the removal efficiency of ammonium nitrogen. The photoautotrophic organisms that this thesis will focus on are eukaryotic algae and prokaryotic cyanobacteria. The overall work will contribute to the development of design criteria for the PAS system, as a simple, yet innovative, technology with low energy requirements, high removal efficiencies of nutrients and organic compounds.

To achieve the objective of this PhD, it is necessary to (i) determine the optimal conditions of the key parameters affecting the interactions within the microalgal-bacterial consortia (SRT, HRT) by assessing the ammonium removal and biomass production of the microalgae-bacteria under different operational conditions and (ii) to determine the key kinetic parameters of the microalgal-bacterial consortia based on laboratory scale experiments and the proposition of a mathematical model. This will allow the understanding of the effects of microalgae on the growth rate of the nitrifying microorganisms and visce versa. Overall, the results will serve as a base to maximize the photo-oxygenation, maximal growth rate and ammonium removal rates when using microalgal-bacterial consortia for nitrogen removal from municipal and high strength wastewater.

3

NITRIFICATION BY MICROALGAL-BACTERIAL CONSORTIA FOR AMMONIUM REMOVAL IN A FLAT PANEL SEQUENCING PHOTOBIOREACTOR

This chapter is based on: Rada-Ariza, María, A., Lopez-Vazquez, C.M., Van der Steen, N.P., Lens, P.N.L., 2017. Nitrification by microalgal-bacterial consortia for ammonium removal in flat panel sequencing batch photo-bioreactors. Bioresource Technology 245, 81-89 https://doi.org/10.1016/j.biortech.2017.08.019

Abstract

Ammonium removal from artificial wastewater by microalgal-bacterial consortia in a flat-panel reactor (FPR1) was compared with a microalgae only flat-panel reactor (FPR2). The microalgal-bacterial consortia removed ammonium at higher rates (100 ± 18 mg $NH_4^+ - N$ L^{-1} d^{-1}) than the microalgae consortia (44 ± 16 mg $NH_4^+ - N$ L^{-1} d^{-1}), when the system achieved a stable performance at a 2 days hydraulic retention time. Nitrifiers present in the microalgae-bacteria consortia increased the ammonium removal: the ammonium removal rate by nitrifiers and by algae in FPR1 was, respectively, 50 (± 18) and 49 (± 22) mg $NH_4^+ - N$ L^{-1} d^{-1}. The ammonium removal by algae was not significantly different between FPR1 and FPR2. The activity of the nitrifiers did not negatively affect the nitrogen uptake by algae, but improved the total ammonium removal rate of FPR1.

3.1 Introduction

Among the widely applied conventional biological nitrogen removal processes, algae-based systems have emerged as an economical solution with high nutrient removal efficiencies (García et al., 2000) and N-recover possibilities (Cai et al., 2013). However, the areal foot-print of algae-based systems needs to be reduced, without compromising effluent quality, while maintaining lower operational costs. These challenges of algae-based systems could be solved by using microalgae-bacteria consortia. Photosynthesis by the algae provides oxygen, which can be used by heterotrophic and ammonium oxidizing bacteria. The carbon dioxide released by carbonaceous oxidation processes can be used by the microalgae

(Subashchandrabose et al., 2011). Moreover, the presence of bacteria within an algal culture improves the settling properties of the biomass (Su et al., 2012b). Better settling properties allow to control the solids retention time (SRT), which permits to operate the system at the shortest optimum SRT. This will promote higher active biomass and higher nutrient removal rates at short hydraulic retention times (HRT) without the risk of biomass wash out (Medina and Neis, 2007; Van Den Hende et al. 2014, 2011; Valigore et al., 2012).

Faster conversion of ammonium either through nitrification or nitritation increases the ammonium removal rates in microalgal-bacterial systems compared to solely algal systems (Rada-Ariza et al., 2015; Wang et al., 2015). There are several factors that affect the growth of algae and bacteria as well as their interactions (Ramanan et al., 2016), such as exchange of micro and macro nutrients, self-shading effect in suspended systems or release of toxins by either bacteria or algae that hinder their mutual growth (Ramanan et al., 2016). These interactions are affected by the operational conditions, such as pH, light intensity, temperature, inorganic carbon competition, nutrients, mixing, dilution rate, SRT, HRT, and algae harvesting (Borowitzka, 1998; Rawat et al., 2011). These parameters can have different optimum ranges for the microbial and microalgal populations. Therefore, it is necessary to study the interactions between the microalgae and bacteria to develop operational guidelines. Through a set of operational guidelines for the SRT and HRT, complete nitrification of ammonium can be achieved by the microalgal-bacterial consortia. Ensuring nitrification is the first step for total nitrogen removal, which can be achieved by applying a further denitrification step.

This study assessed the ammonium removal rates by microalgal-nitrifying (reactor 1, FPR1) and microalgal (reactor 2, FPR2) consortia. The two consortia were cultivated in flat panel photobioreactors operated at 25ºC and pH of 7.5. They were operated at different ammonium loading rates controlled by the influent flow rate, resulting in different HRT. Both consortia were initialy grown under inorganic carbon limitation and subsequently under excess of inorganic carbon. The effect of key operating parameters like SRT and HRT on the ammonium removal rates were evaluated. Furthermore, the nitrogen (ammonium) loading rate (NLR), nutrient removal efficiency, and settleability of the microalgal-nitrifying and microalgal consortia were compared.

3.2 Materials and methods

3.2.1 Reactor set-up

The flat-panel reactors (FPR) used in the experiment are shown in Figure 3.1. They had a total volume of 5.75 L and the dimensions were 0.25m x 0.23m x 0.1 m. They had a heat jacket at the back of the reactor, which was connected to a cooling tower to maintain the desired constant temperature. The influent is entered at the left side of the FPR at the bottom, to ensure full mixing of the synthetic wastewater with the algal (bacterial) biomass. The net working volume was 4 L and the FPRs were operated as sequencing batch reactor (SBR),

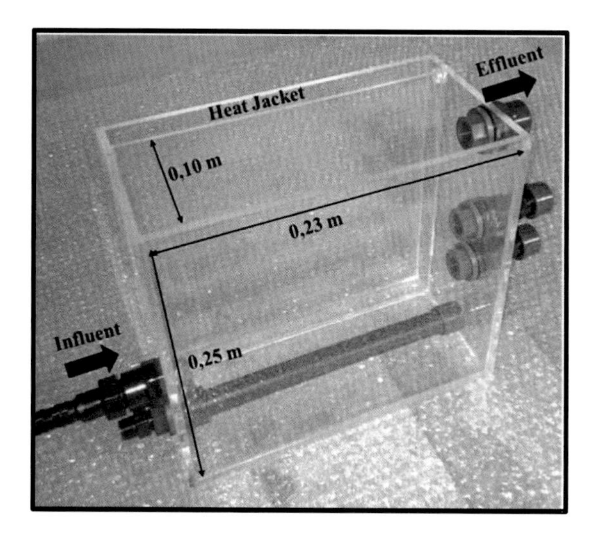

Figure 3.1. Open flat panel reactor (FPR) used in the experiments.

The light intensity on the reactor surface was 700 μmol m^{-2} s^{-1}, and temperature was controlled at about 25 °C using a cooling tower. The FPRs were completely mixed, using magnetic stirrers operated at 500 rpm. The pH in the FPRs was kept around 7.5 by addition of a phosphate buffer solution (PBS) to the synthetic wastewater. The FPRs were operated for 331 d in cycles of 24 h with two feedings per cycle. The two feedings were done in order to divide the nitrogen load and avoid nutrient limited conditions in the FPRs. A 24 hour cycle consisted of: (i) first influent addition (15 min), (ii) first reaction time (11 h 45 min), (iii) second influent addition (15 min), (iv) second reaction time (11h 15 min), (iv) settling (15 minutes), and (v) effluent withdrawal (15 min). The volumes pumped in and withdrawn from the FPRs varied along the periods, and were defined to achieve increasing NLR (Table 3.1).

Table 3.1. Operational conditions and nitrogen (ammonium) loading rate (NLR) in each experimental period applied to the two flat panel reactors.

Period	*Influent ammonium concentration (mg L^{-1})*	*HRT (d)*	***NLR*** *(mg L^{-1} d^{-1})*	*Experimental period length (days)*	*Influent inorganic carbon (Alkalinity) (g L^{-1})*	*Nitrification inhibitor (mg L^{-1})*	*Remarks*
1	27.4	8	3.4	12	0.42	0	Acclimatization periods
2	407.4	8	50.9	7	0.42	0	
3	258.4	8	32.3	12	0.42	0	
4	271.8	4	67.9	23	0.42	0	
5a	283.8	2	141.92	52	0.42	0	Re-inoculation of activated sludge
5b	309.8	2	154.9	62	0.42	0	
5c	297.3	2	148.7	26	3.42	20	Nitrification inhibitor was added in R2 only
5d	299.0	2	149.5	33	3.42	20	
6	268.4	8	33.6	25	3.42	20	
7	253.5	2	126.8	63	3.42	20	

3.2.2 Inoculation

The two FPRs were inoculated with different mixtures of microalgae and bacteria. Reactor 1 (FPR1), set up to develop a microalgal-bacterial consortia, was inoculated with 50 mL containing 10 mL of five different pure cultures of algal strains, and 50 mL of mixed liquor activated sludge from the Harnaschpolder wastewater treatment plant (Delft, The Netherlands). The algae strains used were *Scenedesmus quadricauda, Anabena variabilis, Chlorella sp., Chlorococcus sp. and Spirulina sp.* Their cell density, determined using the Thoma cell counting chamber, was 0.48, 0.09, 0.29, 0.27 and 0.05 cell ml-1, respectively. The activated sludge came from a conventional activated sludge treatment, of which a detailed composition is reported in (Gonzalez-Martinez et al., 2016). The inoculation ratio based on volume was a 1-to-1 ratio of microalgae-to-bacteria. On average, 9 L of mixed liquor activated sludge (3.1 g TSS L^{-1}) was added to FPR1 from day 99 to 122 (3 additions of 1 L each week, during 3 weeks). Reactor 2 (FPR2), assembled as control reactor to enrich a solely microalgal consortia, was inoculated solely with 10 mL each of the five different algae strains.

3.2.3 Composition of the synthetic wastewater

The FPRs were fed with BG-11 medium as synthetic wastewater (Becker, 1994). The nitrogen source was ammonium and the concentration fed to the FPRs was changed throughout the different experimental periods (Table 3.). The phosphorous concentration of the influent remained constant during the FPRs operation (0.08 g L^{-1} of K_2HPO_4). The phosphate buffer used for pH control had a concentration of 0.10 mol of $Na_2H_2PO_4$ and 0.02 mol of NaH_2PO_4. Inorganic carbon was added as

$NaHCO_3$ to supply alkalinity, and concentrations were changed depending on the periods as detailed in (Table 3.1). Organic carbon was not added to the medium since the aim of the experiment was to assess the ammonium removal by nitrifying bacteria in the presence of algae.

3.2.4 Experimental design

Seven periods were studied based on different NLR (Table 3.). The NLR was adjusted with the HRT in order to assess its effect on the microalgal cultures and microalgal-bacterial consortia present in the two FPRs (Table 3.). Alkalinity was another parameter that was varied along the periods.

Periods 1 to 4 were defined for acclimatization of the biomass. Period 5 was subdivided in 4 phases, labelled from (a) to (d). From Period 5a onwards, the HRT was set at 2 days in order to assess the ammonium removal at high NLR, and fresh activated sludge was re-added. Low alkalinity concentrations (0.42 g L^{-1}) were used in Periods 1 to 5b. The alkalinity concentration was increased to 3.42 g L^{-1} from Period 5c onwards. During Period 5d nitrification stopped, therefore, in period 6 the HRT was increased to 8 days to increase the biomass retention. During Period 7, the HRT was again decreased to 2 days. The nitrification inhibitor N-Allylthiourea was added to FPR2 from period 5d until the end of the experiment. The SRT was not controlled but calculated based on the total solid concentrations in the reactor and in the effluent, as described by (Ekama and Wentzel, 2008a).

3.2.5 Sampling and analytical methods

Samples for the determination of $NH_4^+ - N$, $NO_2^- - N$ and $NO_3^- - N$ in the influent and effluent were collected three times per week. Once per week, mixed liquor and effluent samples were collected for the determination of total suspended solids (TSS), volatile suspended solids (VSS) and Chlorophyll-*a* content.

All analytical parameters were determined in accordance to standard methods (APHA, 2005): ammonium and nitrite following the colorimetric method, nitrate using the spectrophotometric method with 2.6-dimethylphenol, and VSS and TSS concentrations by gravimetry. Chlorophyll-*a* was measured using the Dutch standard method NEN-6520. The dissolved oxygen (DO) concentration was measured in-situ and recorded continuously in the two FPRs using a WTW Oxy 3310 electrode (Weilheim, Germany).

3.2.6 Nitrogen balance

The nitrogen mass balance (Appendix A) was calculated to define the nitrogen removal mechanisms. It was assumed that there was no volatilization of ammonium as the pH remained between 7.5 - 8.0 throughout the study periods (Escudero et al., 2014; García et al., 2000; González-Fernández et al., 2011a), and that neither nitrate nor nitrite was consumed by the microalgae (particularly in FPR1), since ammonium concentrations were still left in the effluent. Therefore, the mechanisms of ammonium removal are oxidation by nitrifiers and nitrogen consumption by algae and nitrifiers. The equations used to calculate the nitrogen for each removal mechanism were from Liu and Wang (2012) for partial and full nitrification and Mara (2004) for the algal activity. Based on the removal by each group of microorganisms, the ammonium removal rate (ARR) was calculated. ARR is the amount of ammonium uptake or oxidised per volume of reactor in a specified time by algae and/or nitrifiers. The biomass production of nitrifiers and algae in the FPRs for each cycle was calculated using their nitrogen growth requirements, based on the amount of ammonium oxidised and the amount used for growth by nitrifiers and algae, respectively. The equations applied for the nitrogen growth requirement by nitrifying bacteria and algae are proposed by Ekama and Wentzel (2008) and by Mara (2004), respectively. The detailed calculation are described in SI (Appendix A). Statistical analysis was performed using the t-test (two tailed) in the software Excel.

3.3 Results and discussion

3.3.1 Biomass concentration and production in the FPRs

From Periods 1 to 3, the solids concentration increased in the two FPRs due to the start-up phase in which the biomass has an exponential growth. Further acclimatization to the different operational conditions occurred in FPR1 and FPR2 during periods 3 to 7 (Figure 3.2). After Period 3, the HRT was reduced and the NLR increased by increasing the influent flow rate and effluent discharge volume. The combination of the higher discharge volume and the lower settleability resulted in a decrease in the solids concentration after period 3 (Figure 3.2). Furthermore, from Period 4 (4 d HRT), to Period 5a (2 d HRT), the solids concentration in FPR2 decreased by 30% ($p<0.05$). In contrast, the biomass concentration in FPR1 did not show a significant difference between periods 4 and 5a. Microscopic analysis of the biomass showed that for FPR1 and FPR2 after period 4 until period 7 most of the algae were identified as *Chlorella vulgaris*. This strain is known for its tolerance to high ammonium concentrations (Wang et al., 2010). Also, Cai et al. (2013) reported that *Chlorella vulgaris* has higher nutrient removal efficiencies when compared with other algae strains. Therefore, this species is expected since the synthetic wastewater was rich in ammonium and phosphorous.

The addition of fresh activated sludge in Period 5b increased the solids concentration in FPR1 to 5 g TSS L^{-1}, which later decreased to 1.5 g TSS L^{-1} during the same period. During Period 5c, the solids concentration in the two reactors increased presumably due to the increase in inorganic carbon concentration. Both nitrifiers and algae need inorganic carbon for their growth. The growth of algae increased as more inorganic carbon was available. Mokashi et al., (2016) concluded that 1g L^{-1} of sodium bicarbonate resulted in the highest biomass production and growth rate of the strain *Chlorella vulgaris*. Furthermore, the addition of ATU in the FPR2 stopped nitrification, which increased the availability of ammonium for algae growth. The smaller effluent volume discarded during Period 6, with the purpose of increasing the HRT (8 days), led to a further increase in the solids concentration in the two FPRs. During Period 7, the reduction in HRT decreased the solids concentration in FPR1. In contrast, there was no significant difference between the solids concentrations of Periods 6 and 7 in FPR2 ($p>0.05$).

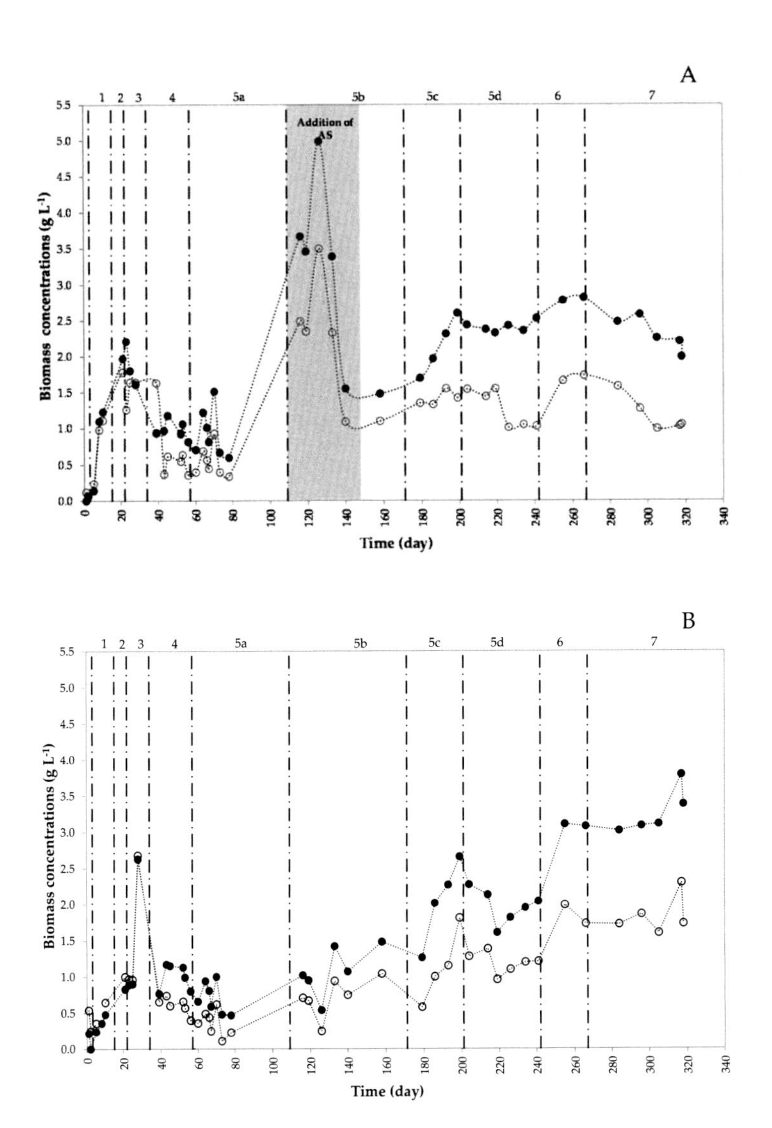

Figure 3.2. Evolution of biomass concentrations in FPR1 (A) and FPR2 (B) along the experimental periods. TSS concentration in the FPR (···●···), and VSS concentration in the reactor (··⊙···).

Overall, both FPRs could not reach a stable operation in the initial Periods 1 to 4 and in parts of Period 5, when analysing the solids concentration. Nevertheless, the FPR stabilized from Period 5d onwards. The biomass production was estimated taking into account the TSS present in the effluent (Figure 3.3), since no biomass was wasted from the reactor. The biomass production can be assumed to be a reflection of the algal and bacterial growth in each period. The biomass growth was higher in FPR1 than in FPR2 from Periods 2 to 4. The high production value for FPR1 in period 5b was due to the AS addition, and is thus not comparable with the other periods.

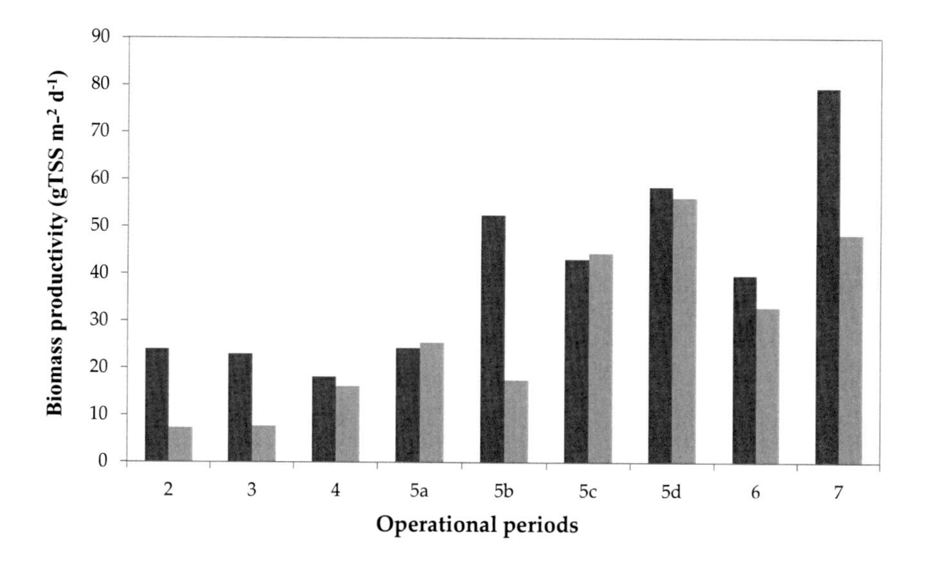

Figure 3.3. Biomass production in FPR1 and FPR2 during periods 2 to 7. FPR1 (■), FPR2 (■).

The highest surface biomass production rate was observed in FPR1 during period 7, with a value of 79.4 g TSS m^{-2} d^{-1} (considering the area perpendicular to the light path of 0.04 m^2). The high solids concentration in the effluent during Period 7 (1.74 ± 0.13 g TSS L^{-1}) for FPR1, and the larger effluent volume discarded of 2 L d^{-1} (compared with Period 6 in which the discarded volume was 1 L d^{-1}) led to an increase in the biomass production during Period 7.

3.3.2 Solids retention time and the effect on ammonium removal rates

The settling time in the sequencing batch operation had the aim to decouple the HRT from the SRT through the retention of biomass in the FPRs. There was no further control of the SRT, thus this parameter depended on the settling properties of the sludge, which determined the TSS retained in the reactor and the TSS lost in the effluent (Table 3.2). The SRT in FPR1 and FPR2 was highly affected by the poor settleability of the microalgal-bacterial biomass and the algal biomass, respectively. From Period 3 to 4, the SRT of the two FPRs decreased due to the increase in the discharged effluent volume, which decreased the HRT from 8d to 4d. The larger discarded volume presumably limited the retention of nitrifiers in FPR1, due to a shorter SRT of 5.0 (± 0.7) days. A further reduction of the HRT to 2 days (Periods 4 to 5a) led to a SRT reduction in FPR1 of 3.5 (± 1.1) days. During these periods, most of the biomass was composed of algae. The addition of fresh activated sludge (AS) in Period 5b, improved the settling properties of the biomass in FPR1, increasing the SRT to 8.0 (± 3.8) days. By the re-addition of AS during this period, a new batch of conventional activated sludge microorganisms such as ordinary heterotrophs, nitrifiers and phosphate accumulating organisms (Gonzalez-Martinez et al. 2016) were added in the reactor, which helped with sedimentation and boosted the nitrification.

Table 3.2. Solids retention time (days) in FPR1 and FPR2 during periods 3 to 7.

Period	*FPR1*	*FPR2*
3	7.0 ± 1.4	18.8 ± 12.3
4	5.0 ± 0.7	5.8 ± 1.8
5a	3.5 ± 1.1	2.6 ± 0.7
5b	8.0 ± 3.8	5.9 ± 2.2
5c	4.2 ± 0.3	4.2 ± 1.1
5d	3.9 ± 0.8	3.5 ± 1.2
6	3.2 ± 0.3	4.3 ± 0.3
7	2.6 ± 0.2	6.2 ± 0.9

For SRTs between 3 and 5 days, 60% of the TSS of the FPRs was lost in the effluent of the reactors. The fraction is calculated using the TSS in the effluent compared with the TSS in the reactor. This percentage illustrates the poor settleability of the biomass which was not expected for FPR1 since the algal-bacterial biomass has proven to increase the settleability properties compared with algal biomass (Su et al., 2012b). However, since organic carbon was not added in the medium, the growth of heterotrophic bacteria was limited in the microalgae-bacteria consortia from FPR1. Furthermore, due to the low biomass retention in FPR1, the conversion rates by the nitrifiers was limited, which negatively affected the ammonium removal within the microalgal-nitrifying FPR1.

The SRTs calculated in this study (Table 3.2), for the different experimental periods were above the minimum SRT (SRT_{min}) for nitrification (2.6 days at 25ºC) calculated based on Ekama and Wentzel, (2008a). However, it was observed that the bacterial biomass did not settle as well as the algae and therefore the SRT for the nitrifiers was probably somewhat lower. On the contrary, if the increase in ammonium loading rate was achieved by increasing the ammonium concentration while keeping the same HRT, this would probably have led to an increment in the growth of nitrifiers, increase of the biomass' settleability thus a longer SRT, which would have resulted in higher ammonium removal rates (ARR). Therefore, the nitrification rates of microalgal-nitrifying consortia and/or micraolgal-bacterial consortia can be increased when a suitable biomass retention is ensured (depending on the environmental conditions). Higher nitrification rates through decoupling of the HRT and the SRT, while operating at a suitable retention time allows to reduce area requirements when cultivating microalgal-bacterial biomass.

The retention of the biomass can be ensured by improving the settling characteristics of the microalgae, which can be achieved when combined with bacteria as reported by several studies (Gutzeit et al., 2005; Medina and Neis, 2007; Lee et al., 2013; de Godos et al., 2014; Van Den Hende et al., 2014). For this, the control of the SRT is extremely important for the development of good settleable microalgal-bacterial biomass (Gutzeit et al., 2005; de Godos et al., 2014). It should be underlined that the properties of microalgae affect the settleability, since their negative surface charge does not favour floc formation. Along the life cycle, algae change their surface charge, with lower surface charges present in older algal cultures (Henderson et al., 2008), which favours agglomeration. Medina and Neis (2007) found that longer SRTs in algae developed a more compact EPS matrix that helped to increase the settleability. In addition, for microalgal-bacterial biomass, factors such as the food/microorganism ratio (F/M) and the wastewater characteristics determine the formation of microalgal-bacterial flocs (Medina and Neis, 2007; Van Den Hende et al., 2014). For instance, long retention times (SRT 40 days) resulted in improved settleability of microalgal-bacterial biomass

(Valigore et al., 2012). Depending on the total organic carbon to total inorganic carbon ratios in the wastewater, flocs of microalgae and bacteria (heterotrophic and autotrophic) could be dominated either by bacteria or microalgae (van den Hende et al., 2014). During the experiment the short SRTs and HRTs resulted in an increase in dispersed algae concentrations as observed during the first phases (1 - 5a). Thus, considering that short SRTs and HRTs favour rapid algal growth, the formation of flocs will be hard to achieve as microalgae present a negative surface charge and form a less compact matrix of EPS than bacteria. Thus, loss of bacterial biomass becomes a risk. Therefore, to maximize floc formation, stimulation of growth of heterotrophic biomass is recommended. This can help to improve settleability while achieving longer SRTs, increased EPS formation will eventually reduce the surface charge of the algae and thus result in an overall increase of the biomass retention and ammonium conversion rates.

Operational strategies such as maintaining the SRTs>>HRTs in the initial periods can be implemented in a microalgal-bacterial reactor and HRAP to ensure good biomass settleability, yet it is important to start with a biomass that has a minimum settleability. This can be achieved by addition of activated sludge during the start-up period, which in overall can trigger algal-bacterial flocs formation. Rada-Ariza et al. (2015) tested several SRTs (1 - 15 days) with a HRT of 1 day in a continuous flow microalgal-bacterial system. A long SRT in the initial periods was achieved by recirculation of the biomass. This together with a low HRT, helped to develop a well settleable biomass. The good settleability allowed to further decrease the SRT, achieving the highest removal rate at an SRT of 3 day (0.075 $\pm$ 0.002 g $NH_4^+ - N$ L^{-1} d^{-1}). By maintaining an optimum balance of solids through

SRT and HRT strategies, microalgal-bacterial systems in HRAP can achieve high removal rates treating ammonium-rich wastewater, without any external aeration or increase of their aerial footprint.

3.3.3 Fate of nitrogen in the FPRs

During the acclimatization Periods 1 to 4, the concentration of nitrogenous compounds in the effluent showed similar trends in both FPRs (Figure 3.4). Ammonium was removed to negligible values by the sole action of algae, since no nitrate or nitrite was measured in FPR1 and FPR2 and because denitrification was unlikely due to several reasons: (i) high DO values above oxygen saturation, (ii) no external addition of organic carbon, and (iii) not long enough SRTs. Since the pH oscillated between 7.5 to 8.0 and the temperature was kept at 25^{0}C, ammonium stripping was ruled out as a potential removal pathway. The ARR for FPR1 during the acclimatization periods ranged between 3.9 (± 0.2) to 74.8 (± 4.4) mg $NH_4^+ - N$ L^{-1} d^{-1}, and were not significantly different from those observed in FPR2 (with values between 3.9 (± 0.3) - 73.3 (± 3.3) mg $NH_4^+ - N$ L^{-1} d^{-1}) (Figure 3.5).

During Period 5a, the HRT was decreased to 2 days and the NLR increased accordingly to supply more ammonium to avoid nitrogen limitation, and increase nitrification rates. Due to the limitation of algae to remove such nitrogen loads in a relatively short period of time (in FPR1 and FPR2), and nitrification was not yet occurring in FPR1 as intended, the ammonium concentrations increased in the effluent of both reactors (FPR1 and FPR2). The ARR by algae decreased to 34.3 (± 26.8) and 28.6 (± 25.7) mg $NH_4^+ - N$ L^{-1} d^{-1} in FPR1 and FPR2, respectively (Figure 3.5).

Nitrification started in Period 5b in the FPRs. The highest nitrate concentration of 41.3 (± 8.6) mg $NO_3^- - N$ L^{-1} was observed in FPR1, while in FPR2, nitrate concentrations were around 12.7 (± 7.9) mg $NO_3^- - N$ L^{-1} (Figure 3.4). The addition of AS and possibly growth of nitrifiers increased the nitrification rates. The ARR increased in period 5b up to 72.9 (± 36.9) and 50.2 (± 33.5) mg $NH_4^+ - N$ L^{-1} d^{-1} in FPR1 and FPR2, respectively (Figure 3.5). The ARR in FPR1 was significantly higher ($p < 0.05$) than in FPR2, which is attributed to the combined removal by algae and bacteria.

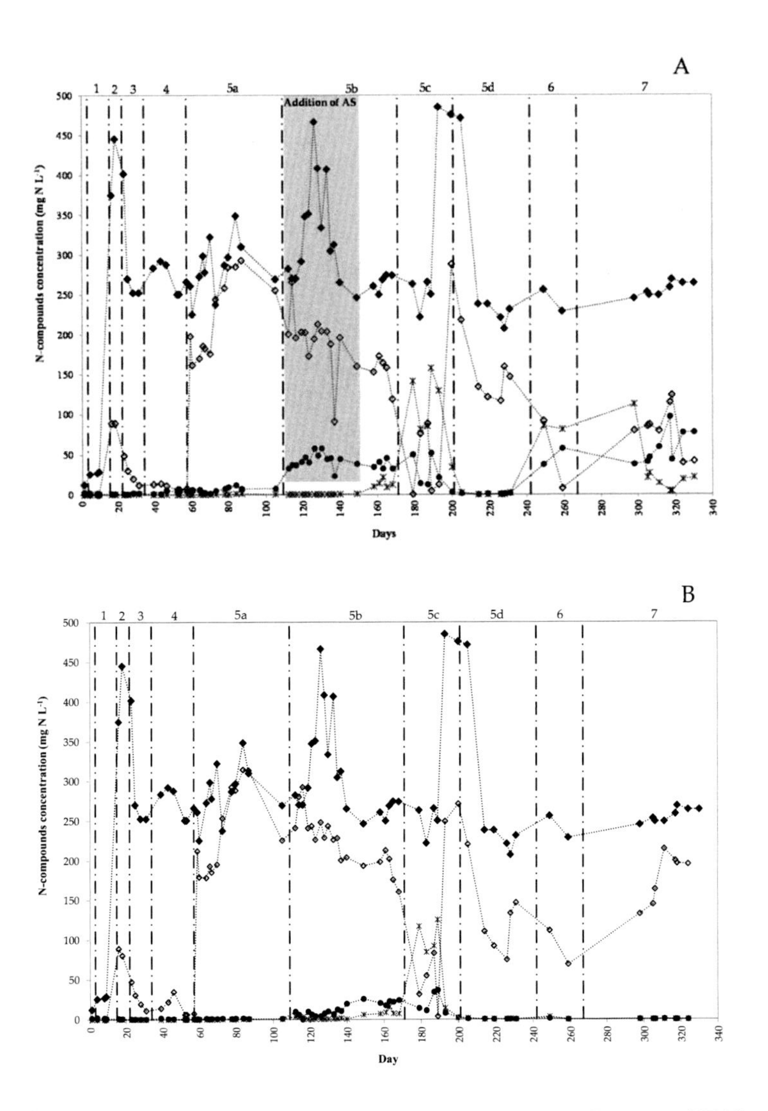

Figure 3.4. Concentrations of nitrogen compounds in the effluents of FPR1 (A), and FPR2 (B), along the experimental periods. Legend: (····◇····) effluent NH_4^+-N, (····✱····) effluent NO_2^--N, (····●····) effluent NO_3^--N, and (····◆····) influent NH_4^+-N.

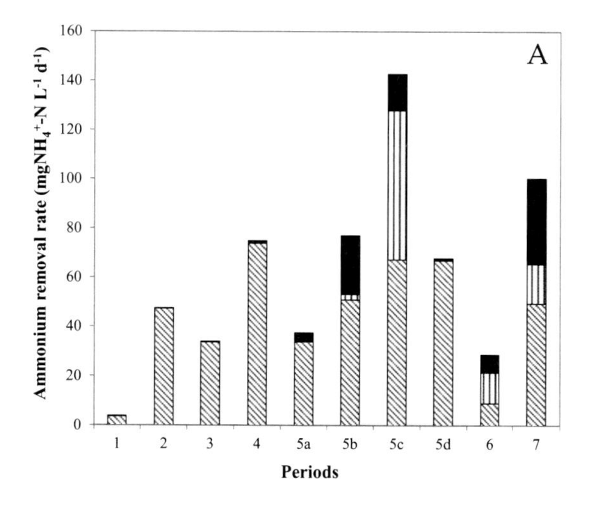

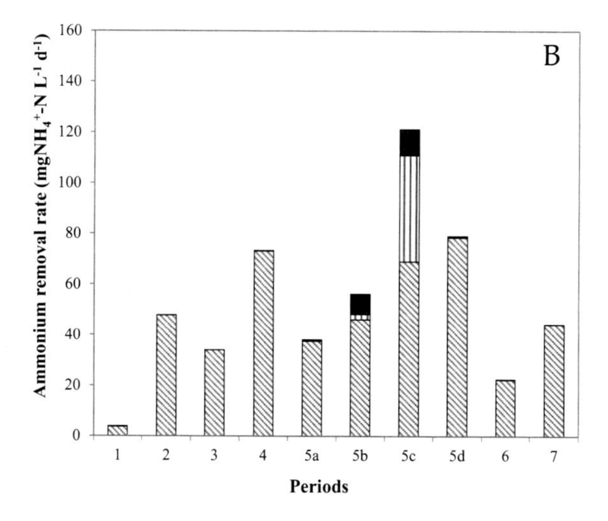

Figure 3.5. Total ammonium removal rates for nitrifiers (AOB & NOB) and algae based on the nitrogen balance in FPR1 (A) and FPR2 (B) during the different operational periods. NOB (■), AOB (◫) and algae (▨)

From period 5c onwards the alkalinity concentration was increased from 0.42 to 3.4 g L^{-1} $NaHCO_3$ in order to enhance the nitrification rates. The increase in alkalinity (period 5c) boosted the rates of ammonium oxidation. The ammonium concentration in the effluent (period 5c) in both FPR decreased compared with Periods 5a and 5b ($p < 0.05$). The ARR in the two reactors increased on average by 57 mg $NH_4^+ - N$ L^{-1} d^{-1}. However, during Period 5C in FPR2 the measurement of nitrite production confirmed the growth of AOB, therefore, ATU was added from day 190 onwards in FPR2 to inhibit nitritation (Figure 3.4).

The increase in inorganic carbon in the medium during Period 5C boosted the ARR by both algae uptake and nitrification, achieving the maximum rate in FPR1 of 143($\pm$ 6) mg $NH_4^+ - N$ L^{-1} d^{-1}. Before this period, the reactors had an alkalinity limitation. In Periods 5c to 7, for the influent ammonium concentration, the amount of alkalinity required for nitrification, assuming that 80% of the ammonium would be nitrified, was 1.9 g $NaHCO^-_3$ L^{-1}, while photosynthesis required 0.4 g $NaHCO^-_3$ L^{-1}. Therefore, alkalinity in the influent was not a limiting factor (3.4 g $NaHCO^-_3$ L^{-1}) from Period 5b onwards in the FPR1. Since both photosynthesis and ammonium oxidation require inorganic carbon, alkalinity must be sufficient in the wastewater to ensure efficient and faster ammonium removal.

Following the alkalinity increase, the ammonium oxidising bacteria (AOB) activity increased resulting in a maximum effluent concentration of 105.1 ($\pm$ 46.0) mg $NO_2^- - N$ L^{-1} in FPR1, with no nitrate production. This shows that the nitrite oxidising bacteria (NOB) activity was still limited and did not further oxidize such nitrite concentrations to nitrate. Likely, this was due to the high concentrations of ammonium in the influent during Period 5c, which corresponds to an ammonia concentration of 10.7 mg NH_3 L^{-1}. This value falls in the low range for AOB inhibition defined by Anthonisen et al. (1976) (10 - 150 mg NH_3 L^{-1}), and more importantly, it is considerably higher than the maximum NOB inhibition range of 0.1 - 1.0 mg NH_3 L^{-1}. Possibly, these high concentrations also contributed to the inhibition of NOB and AOB by free nitrous acid (FNA). In relation with nitrite, the average nitrite concentration in FPR1 during Period 5C corresponds to 3.3 mg HNO_2 L^{-1}, which is above the high range proposed by Anthonisen et al. (1976) of 0.2 - 2.8 mg HNO_2 L^{-1} for both AOB and NOB inhibition. Ultimately, these concentrations lead to an inhibition of the activity of the nitrifiers in FPR1.

Ammonia toxicity may prevail at low HRT and high ammonium loading rate, especially for NOB and algal species. Several studies have provided different limiting concentrations for ammonium. Toxicity of ammonia to algae was reported by Azov and Goldman (1982) with a 50% reduction in photosynthesis at concentrations of 22.1 mg NH_3-N L^{-1} between pH 8 - 9.5. Tuantet et al. (2013) reported algae inhibition at 140 mg NH_3-N L^{-1} at pH 8.2. He et al. (2013) observed a decrease in algal growth when the concentrations of ammonium increased in the influent from a maximum observed growth rate (μ_{mobs_algae}) of 0.92 d^{-1} at 30 mg NH_4^+-N L^{-1} to 0.33 d^{-1} at 143 mg NH_4^+-N L^{-1} (at pH between 6.96 to 7.10). A neutral pH in an algal system is for optimum nutrients removal necessary (Liang et al., 2013). Similarly, for nitrifiers, pH fluctuations can cause nitrite accumulation, as a stable pH avoids shifts in the ammonium/ammonia, as well as the nitrous acid equilibrium.

In order to recover the system, a smaller volume was discharged from the FPR during Period 6, to increase the apparent HRT and reduce the NLR. This action benefited the nitrifiers and algae in FPR1, since nitrate and nitrite concentrations in the effluent were on average 47.8 (± 14.1) mg $NO_3^- - N$ L^{-1} and 83.8 (± 3.1) mg $NO_2^- - N$ L^{-1}, respectively, and ammonium concentrations decreased in the effluent (7.6 mg $NH_4^+ - N$ L^{-1}).

Once nitrification was restored, the HRT was reduced to 2 days (period 7), and nitrification continued. The ARR reached 100.2 (± 17.9) mg $NH_4^+ - N$ L^{-1} d^{-1}, this corresponds to a surface removal rate of 10.2 g $NH_4^+ - N$ m^{-2} d^{-1}. This value is higher than 2.0 g $NH_4^+ - N$ m^{-2} d^{-1} reported by Sutherland et al. (2014) for a pilot high rate algae pond (HRAP) operated at a 4 days HRT, and close to the value reported by Godos et al. (2009) of 6.7 g$NH_4^+ - N$ m^{-2} d^{-1} for a pilot HRAP treating piggery wastewater at a HRT of 10 days, where the main removal mechanisms was nitrification. The ARR of FPR1 is significantly higher ($p < 0.05$) than that of FPR2 (Figure 3.5). Since no other removal mechanism could take place, it is assumed that the action of nitrifiers in FPR1 contributed to the doubling of the ARR compared with the ammonium removal activity observed in FPR2 driven solely by the activity of microalgae.

3.3.4 Total and specific ammonium removal rates by algae and nitrifiers in the FPRs

Total ammonium removal rates in FPRs

Ammonium removal rates by nitrifiers and algae were calculated based on the nitrogen balance of the FPRs. Figure 3.5 shows the contribution of the algae, AOB and NOB (nitrifiers) to the total ammonium removal rate. In the first four periods, the daily nitrogen removal by the algae did not differ between the two FPRs (p >

0.05), and the oxidation of ammonium was not detected. Likewise, the total ARR by the FPRs was similar in the two reactors ($p > 0.05$) during the first 4 periods.

When nitrification took place, the total ARR by both algae and bacteria in FPR1 (Figure 3.5) was higher than in FPR2 ($p < 0.05$). In FPR1, during the periods in which nitrification ocurred (5b, 5c, 6 and 7), the ARR by algae ranged between 0.4 – 2.8 mg $NH_4^+ - N$ L^{-1} h^{-1}, while in FPR2 it was between 0.8 - 4.0 mg $NH_4^+ - N$ L^{-1} h^{-1}. Furthermore, the average ARR by algae in Periods 5a to 7 (except Period 6) between FPR1 and FPR2 (FPR1 = 53.6 (± 13.9) and FPR2 = 55.0 (± 17.6) mg $NH_4^+ - N$ L^{-1} h^{-1}) are not significantly different ($p > 0.05$). The presence of nitrifiers in the system thus did not inhibit the ammonium consumption by the algae, neither the possible shading effect of nitrifiers on algae. On the contrary, the presence of nitrifies improved the total ammonium removal rates of FPR1.

In a laboratory study using an algal-bacterial consortia, Karya et al. (2013) found that nitrification can reach up to 7.7 (± 4.4) mg $NH_4^+ - N$ L^{-1} h^{-1}. This value is higher than the ones obtained in this research. Presumably, this was due to the low biomass retention in the FPRs (Table 3.2). Despite having obtained relatively higher total ammonium removal efficiencies in the periods in which nitrification occurred, the maximum rate achieved was 5.9 mg $NH_4^+ - N$ L^{-1} h^{-1} (Period 5c), while higher rates of 14 - 21 mg $NH_4^+ - N$ L^{-1} h^{-1} can be achieved in activated sludge (Azimi and Horan, 1991).

Specific ammonium removal rates by nitrifiers and nitrogen uptake by algae

The nitrogen balance was used to estimate the nitrifying biomass and nitrogen uptake by algae, the equations used and detailed calculations can be seen in

Appendix A. The nitrogen uptake by algae was calculated based on the nitrogen balance and compared with theoretical values calculated using the stoichiometric yield of ammonium consumption per algal biomass formed proposed by (Mara, 2004) of 9.2%, and the TSS and SRT of the FPR1. However, when comparing these values it is found that the theoretical values are lower than the ammonium uptake by algae based on nitrogen balance. For instance, during Period 7, in FPR1 using the photosynthesis expression proposed by (Mara, 2004) , algae take up 7.4 mg $NH_4^+ - N$ h^{-1} in FPR1, and 4.4 mg $NH_4^+ - N$ h^{-1} in FPR2, while with the calculations based on the nitrogen balance algae in FPR1 consume 7.1 (±3.2) and in FPR2 6.4 (±2.2) mg $NH_4^+ - N$ h^{-1}. In order to obtain similar values, the nitrogen content in algal biomass must be between 9 - 13%. Therefore, the nitrogen uptake by algae per gram of biomass formed in the reactors slightly exceeds the 9.2% proposed by Mara (2004). Ruiz et al. (2011), through a set of batch tests using *Chlorella vulgaris* at different nitrogen concentrations (5.8 - 226.8 mg $NH_4^+ - N$ L^{-1}), obtained percentages of nitrogen in the biomass between 11.5 - 21.8%. The higher uptake of nitrogen by algae can be attribute to storage of N within the algae cell. Further studies are necessary to assess how and under which conditions the algal biomass can store N-compounds in addition to the uptake required for growth.

Using the values of the nitrogen mass balance and the nitrifying biomass, the specific ammonium removal rates were calculated. The specific removal rates by nitrifiers for ammonium were 0.9, 1.6, 0.2 and 1.3 g $NH_4^+ - N$ $gVSS_{nitrifiers}^{-1}$ d^{-1}, for Periods 5b, 5c, 6 and 7, respectively. Karya et al. (2013) reported a value of 1.4 g $NH_4^+ - N$ $gVSS_{nitrifiers}^{-1}$ d^{-1}. This value is comparable with the one obtained in Periods 5c and 7, in which the highest nitrification rates were achieved. When compared with an optimized activated sludge system (4.5 g $NH_4^+ - N$ g $VSS_{nitrifiers}^{-1}$ d^{-1}) (Ekama and Wentzel, 2008a), the values obtained in the FPR1 are significantly lower. This can be related to the low SRTs between 3 - 4.0 days (Table 3.2), which was very close to the minimum SRT for nitrification in FPR1 (2.6 days).

3.3.5 Implications of using microalgal-bacterial consortia for ammonium removal

In algal systems, area reduction is one of the key challenges to face, as the need for higher nutrient removal efficiencies will lead to a higher area requirement. Using the total ammonium removal rates obtained in this research during Period 7 and using a depth of 0.1 m, the ammonium removed per unit of area is 4.4 g $NH_4^+ - N$ m^{-2} d^{-1} for FPR2 (microalgal consortia), while it was 10.2 g $NH_4^+ - N$ m^{-2} d^{-1} for FPR1. This higher value in FPR1 can be attributed to the activity of nitrifiers, which was the main reason for higher ammonium removal rates in comparison to FPR2. This allowed to reduce the area requirements, while operating at high ammonium loading rates, e.g. 126 mg $NH_4^+ - N$ L^{-1} d^{-1}, and lower HRT, e.g. 2 days. Despite these promising values, it must be noted that these calculations and estimations may change when upscaling the technology to pilot or full scale.

When comparing the removal rates per unit of area with other studies, it is observed that (Karya et al., 2013) obtained a value of 10.5 g NH_4^+ m^{-2} d^{-1} (removal rate: 0.185 g $NH_4^+ - N$ L^{-1} d^{-1} and depth of 0.057 m) using a microalgal-bacterial consortia, which is similar to the value obtained for FPR1 in this study. At laboratory scale, using urine as influent medium, (Tuantet et al., 2014) achieved higher ammonium removal rates of up to 1.3 g $NH_4^+ - N$ L^{-1} d^{-1} with *Chlorella sorokiniana* cultured in a short-light path flat photobioreactor (5 mm width) at an HRT of 1 day. The values provided by Tuantet (2015) were used to compare the microalgal-bacterial consortia removal rate per unit of area with a single algal strain system. In that system, the rate of nitrogen removed per area would be 6.5 g $NH_4^+ - N$ m^{-2} d^{-1}, which is higher than the value obtained for FPR2, but lower than the value for FPR1. Furthermore, the total ammonium removal rate reported by (Tuantet et al., 2014) is noticeably higher than the ones obtained in this research (Figure 3.5), possibly due to the short light path (5 mm) which benefitted the light penetration. In practice, light paths will be longer, which will result in higher area requirements. For more practical situations, Park and Craggs (2011) reported an ammonium removal rate of 0.16 mg $NH_4^+ - N$ L^{-1} h^{-1} in a pilot HRAP treating real domestic wastewater. Based on this rate reported and the depth of the pilot HRAP (0.3 m), the surface removal rate of ammonium is estimated to be 1.1 g $NH_4^+ - N$ m^{-2} d^{-1}, which is noticeably lower than the values obtained in this research.

Ensuring higher nitrification rates in microalgal-bacterial consortia is the first step to ensure total nitrogen removal. HRAP using microalgal-bacterial consortia should ensure the decoupling between the SRT and the HRT. It is important that the SRT is higher than the minimum SRT required for nitrification. Thus SRT control through biomass wasting and/or development of a good settleable biomass to avoid wash out it is important for a stable operation. Once nitrification is achieved, denitrification can take place (de Godos et al., 2014; Wang et al., 2015). This can be achieved by introducing dark periods in the operational cycles and ensuring sufficient organic carbon. Thus, a full nitrification-denitrification microalgal-bacterial consortia treatment can be implemented as a secondary treatment for an anaerobic digestion effluent.

3.4 Conclusions

A microalgal-bacterial consortium in a flat-panel photobioreactor removed ammonium from artificial wastewater at higher rates (100±18 mg $NH_4^+ - N$ L^{-1} d^{-1}) than an algae-only system (44±16 mg $NH_4^+ - N$ L^{-1} d^{-1}) at an HRT of 2 days. Nitrification was the mechanism that caused the increase in ammonium removal. This only occurred when the growth and retention of biomass was sufficient to achieve an SRT higher than the minimum SRT for nitrifiers. Consequently, control of the SRT and HRT is key to increase the nitrification rates in microalgal-bacterial systems.

4

AMMONIUM REMOVAL MECHANISMS IN A MICROALGAL-BACTERIAL SEQUENCING-BATCH PHOTOBIOREACTOR AT DIFFERENT SRT

This chapter is based on: Rada-Ariza, María, A., Lopez-Vazquez, C.M., Van der Steen, N.P., Lens, P.N.L.. Ammonium removal mechanisms in a microalgal-bacterial sequencing-batch photobioreactor at different SRT. Algal Research (Submitted)

Abstract

Microalgal-bacterial consortia have important advantages over conventional activated sludge systems by achieving full nitrification and organic carbon oxidation without the need of external oxygen supply. This study assessed the different ammonium removal mechanisms and oxygen production of a microalgal-bacterial consortium at the different solids retention times (SRT) of 52, 48, 26 and 17 days treating synthetic wastewater. The ammonium removal efficiency exceeded 94%, while the total nitrogen removal efficiency was higher than 70% at the different SRTs applied. The main nitrogen removal mechanism was through nitrification/denitrification, followed by algal cell synthesis and bacterial nitrogen growth requirements. Shorter SRTs favoured the nitrification/denitrification processes over the assimilation of nitrogen by algae. The highest volumetric ammonium removal rate observed was 2.12 mgNH_4^+-N L^{-1} h^{-1} at an SRT of 17 d. The total gross oxygen production at the different SRTs ranged between 0.2 and 0.3 kg O_2 m^{-3} d^{-1}, reaching highest production at a 52 d SRT. The differences in oxygen production between the different SRTs are attributed to the algal biomass content and light attenuation. The oxygen consumption decreased at shorter SRTs due to a decrease in the respiration of the microalgal-bacterial biomass. This study showed that the SRT is a key operational parameter that allows to control the nutrient removal processes and observed growth of the microalgal-bacterial consortia.

4.1 INTRODUCTION

The conventional activated sludge process is a widespread technology for wastewater treatment. Artificial aeration can account for between 45 to 75% of the energy consumption in activated sludge plants (Rosso et al., 2008; Lee et al., 2015; Fan et al., 2017). During the last years more attention has been paid to algae-based systems as an alternative technology to the high energy consuming conventional wastewater treatment systems (de Godos et al., 2014; van den Hende et al., 2014b; van der Steen et al., 2015; Rada-Ariza et al., 2017) . Algae-based systems are natural and sustainable technologies to supply oxygen through photosynthesis by making use of the autotrophic metabolism of microalgae and cyanobacteria that utilize light to produce oxygen (Subashchandrabose et al., 2011).

High rate algae ponds (HRAP) have emerged as optimized or re-engineered waste stabilization ponds (WSP). HRAP are designed to operate at higher loading rates and shorter retention times, shallower depths and higher algal productivities (Evans et al., 2005) than conventional WSP. HRAPs often favour the development of microalgal-bacterial consortia that can reduce the concentrations of pollutants through the dual action of microalgae and bacteria. Ammonium removal rates between 0.12 and 5.6 mg NH_4^+-N L^{-1} h^{-1} have been reported in several HRAP studies with algal-bacterial consortia (Evans et al., 2005; García et al., 2006; Park and Craggs, 2011; Wang et al., 2015). Since microalgae and bacteria have shown promising results for nutrient-rich wastewaters, there is an increasing need and interest to further develop the 'photo-activated sludge' systems (van der Steen et al., 2015). At lab-scale, certain studies have reported an ammonium removal efficiency and rate in a sequencing batch photo-bioreactor of up to 85% and 7.7 mg NH_4^+-N L^{-1} h^{-1}, respectively, using synthetic wastewater with an influent concentration of 50 mg NH_4^+-N L^{-1} (Karya et al., 2013). Other researchers have achieved a maximum removal rate of 4.1 (± 0.7) mg NH_4^+-N L^{-1} h^{-1} and ammonium removal efficiency of 70% in a sequencing batch microalgal-bacterial system fed with synthetic wastewater with an influent concentrations of 250 mg NH_4^+-N L^{-1} (Rada-Ariza et al., 2017), while in other studies removal rates of up to 0.13 mg NH_4^+-N L^{-1} h^{-1} in a pilot-scale HRAP enriched with a microalgal-bacterial consortia treating pikeperch wastewater (32.7 ± 11.7 to 68.4 ± 12.4 g L^{-1} of total nitrogen) have been reported by van den Hende et al. (2014a).

The use of microalgal-bacterial systems presents several challenges in their operation, since the reactors need to be designed to carry out bacterial and algal processes efficiently. Therefore, clear specifications for operational conditions need to be studied and researched, in order to improve bioflocculation, harvesting and biomass control (Muñoz and Guieysse, 2006). Gutzeit et al. (2005) showed an ammonium removal efficiency of 60% in a pilot scale system using microalgal-bacterial biomass treating pre-treated sewage (53.9 ± 7.6 mg NH_4^+-N L^{-1}) operated at an SRT of 40 days. Furthermore, Gutzeit et al. (2005) based on the pilot-scale and laboratory results proposed operational conditions for SRT and hydraulic retention time (HRT) between 20 – 25 days and 2 - 3 days, respectively. On the other hand, Arashiro et al. (2016) reported ammonium removal efficiencies up to 98% in a laboratory scale photobioreactor treating the centrate from an anaerobic digestor treating swine manure (236 ± 19 mg NH_4^+-N L^{-1} at an SRT of 7 and 11 days. Therefore, shorter SRTs on microalgal-bacterial systems can be achieved and still achieve high removal efficiencies. Furthermore, to our knowledge few studies combined the optimization of both: (i) increase of the ammonium removal rates and (ii) optimum operational for biomass retention on microalgal-bacterial systems. In addition, the effect of the biomass retention on the different processes within a microalgal-bacterial system treating nitrogen-rich wastewater needs to be analysed.

Focusing on the operational conditions usually implemented on algal systems, it is not common to define the sludge retention time (SRT), since usually there is no biomass recycling in open or closed algae reactors and because the common practice is to harvest the biomass for further uses. Nevertheless, when working with a consortium of microalgae and activated sludge bacteria for nutrient and organic matter removal through photo-oxygenation, the SRT plays an important role. In fact, it is the most important design and operating parameter of activated sludge systems (Ekama and Wentzel, 2008a). Furthermore, biomass retention is necessary in microalgal-bacterial reactors to ensure higher nitrification rates (Rada-Ariza et al., 2017, 2015). In addition, the effects of light attenuation caused by the solids concentration, which is strongly dependent on the SRT, and its effects on the ammonium removal rates have also been addressed elsewhere (Arashiro et al., 2016). However, there is a lack of comprehensive studies that evaluates the combined effects of these key operating conditions. The present study was carried out with the aim of optimizing key operational conditions affecting the microalgal-bacterial consortium in a sequencing batch photobioreactor. Therefore, this paper assesses the effects of different SRTs on the ammonium removal, oxygen production and biomass productivity of a microalgal-bacterial consortium.

4.2 MATERIALS AND METHODS

4.2.1 Photobioreactor set-up

The cylindrical glass jacketed reactor (internal diameter 11.5 cm), described in detail by Karya et al. (2013), had a capacity of 1L, the light intensity and temperature were 25.9 µmol m^{-2} s^{-1} and 28 ºC, respectively, temperature was maintain constant using a cooling tower which was connected to the reactor. The light intensity was measured at the inner surface of the reactor wall using a Photometer model Li-250 (Li-COR, United States). The light source was provided by four white lamps (40W, Phillips, The Netherlands) positioned around the reactor. pH in the reactor was maintained at a constant value of 7.5 using NaOH and HCl solutions which had a concentration of 0.2 M each. The dosage of the solutions was controlled and the pH set point (7.5) was controlled using a Bio-Console Applikon Holland system. The reactor was operated as a sequencing batch reactor (SBR) consisting of two cycles of 12 hours per day, and the total duration of the experiment was 310 days. The influent was 1 L per day and the volume discarded per day was 1 L, divided in 0.5 L in each cycle every 12 hours.

The cycles had two different operational schemes (Figure 4.1). In the first period (1), the SRT was 48 days and the cycles consisted of filling, followed by an aerobic phase, anoxic phase, second aerobic phase, settling and effluent withdrawal. The aerobic phase was under constant illumination for a total of 7.5 hours. For this period, the synthetic wastewater, containing nitrogen (N), phosphorous (P) and trace elements, was fed at the beginning of the aerobic phase, while the organic carbon source (acetate) was added at the beginning of the dark phase. In the subsequent periods (2A, 2B and 2C), the SRT was gradually adjusted to 52, 26, and 17 days, respectively. The cycle scheme was different than the one applied in period 1, and started with a filling phase, followed by an anoxic phase, aerobic phase, second anoxic phase (during which there was the second filling phase), second aerobic phase, settling and effluent withdrawal. During this cycle scheme the aerobic phases were under constant illumination for a total of 8.5 hours. From period 2A onwards, the artificial wastewater contained both the organic carbon source and the required N, P and trace elements to simulate the composition of municipal wastewater. In the experimental periods 2A, 2B and 2C, there were two feedings of artificial wastewater at the beginning of the anoxic phases, both containing ammonium and organic carbon. In order to prevent the development of biofilm on the reactor walls, the reactor was cleaned twice a week. The control of the SRT in the reactor was done through the waste of the biomass as recommended elsewhere (Rada-Ariza et al., 2017). For the determination of the actual SRT (Valigore et al., 2012), the concentration of the wasted biomass and the solids concentration in the effluent was taken into account in accordance to the following equation:

$$\boldsymbol{SRT} = \frac{V_R}{Q_W + Q_S\left({X_S}/{X_R}\right)} \tag{4.1}$$

Where:

V_R= Reactor volume (L)

Q_W= Flow rate of the waste of sludge (L d^{-1})

Q_S= Effluent flow rate (L d^{-1})

X_S= TSS concentration in the effluent (g L^{-1})

X_R= TSS concentration in the reactor (g L^{-1})

Therefore, in order to determine the amount of sludge waste per day, the measurements of the TSS in the effluent in the reactor were done. Secondly, the SRT desired was defined, and then using the equation *(4.1)*, the Q_W was determined. As soon as the change in the sludge waste was done, the first weeks the solids concentrations were measured in order to verify that the SRT was achieved.

4.2.2 Growth medium, microalgal-bacterial consortia and inoculation

Artificial wastewater was used as growth medium for the microalgal-bacterial consortia. The composition of the artificial wastewater was a modification of the BG-11 medium (Becker, 1994): ammonium was adjusted to 23 mg NH_4^+-N L^{-1} and sodium acetate was used as the carbon source at a concentration of 200 mg COD L^{-1}. The biomass was composed of a mixture of microalgae, nitrifiers and ordinary heterotrophic bacteria (OHO). In order to start with a biomass that contained a large population of nitrifiers (ammonium and nitrite oxidizing bacteria (AOBs and NOBs)), a reactor previously cultivated with activated sludge from Harnaschpolder wastewater treatment plant (Delft, The Netherlands) as initial inoculum was fed with ammonium and other trace elements (no organic carbon) to enrich the nitrifying bacteria.

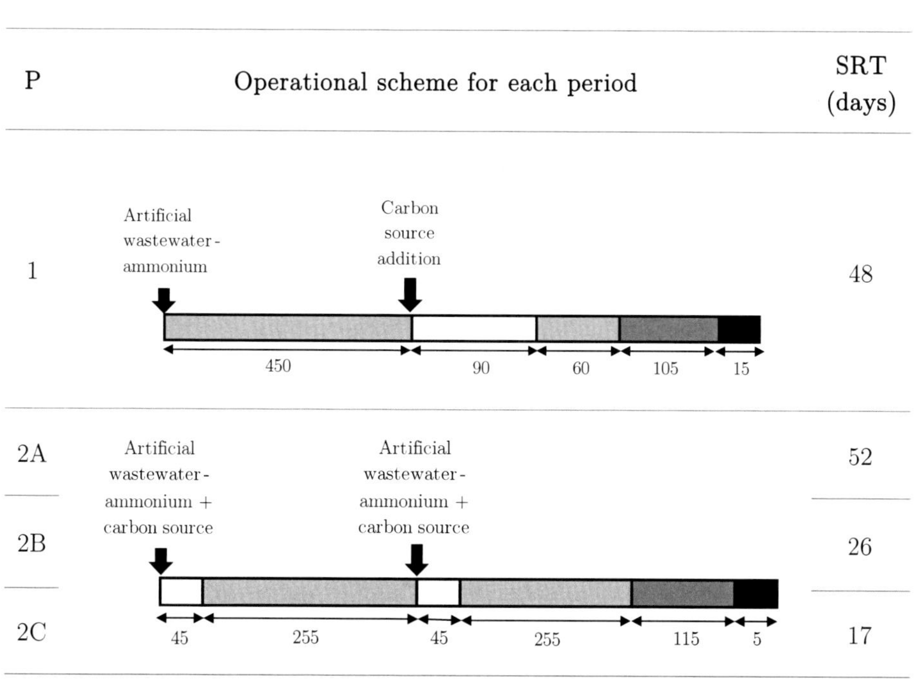

Figure 4.1. Operational scheme, composition of the synthetic wastewater and SRT's applied in the different operational periods assessed in this study. The duration of each phase is presented in minutes below each scheme. P: Period, aerobic phase (▪), anoxic phase (☐), settling phase (▪), and effluent withdrawal phase (■).

The inoculation of the photo-bioreactor was carried out at different days: on day 1, 100 ml of sludge rich in AOBs and NOBs was added. The reported total suspended solids (TSS) of the biomass used as inoculum was 2.5 g L^{-1}. Additionally, 50 ml of fresh activated sludge from Harnaschpolder wastewater treatment were also added. On the 8th day, 5 pure cultures of algae species (total 50 ml) were inoculated: *Scenedesmus quadricauda*, *Anabaena variabilis*, *Chlorella* sp., *Chlorococcus* sp., *Spirulina* sp., as well as unidentified algae from a canal in Delft. Prior to addition, the concentration of chlorophyll-*a* in the algae mixture was 3 mg L^{-1}. After the 8th day, the biomass had a TSS concentration of 2.1 g L^{-1}, with a volatile suspended solids (VSS) and chlorophyll-*a* concentration of 1.1 g L^{-1} and 10.8 mg L^{-1}, respectively.

4.2.3 Sampling and analytical methods

Samples from the influent and effluent (taken at the end of the withdrawal stage) were analysed daily for nitrogenous compounds and COD. Additional samples were collected daily at the end of the second filling time to determine the chlorophyll-*a*, VSS and TSS concentrations.

In each of the periods (1, 2A, 2B, and 2C), certain cycles were analysed in detail through collection of different samples every half an hour for the determination of nitrogenous compounds and COD concentrations. A total of 4 cycles for periods 1 and 2C; 3 cycles for period 2A; and 5 cycles for period 2B were analysed. The detailed data collected was used to estimate the nitrogen and oxygen balances, as well as for biomass characterization and removal rates.

Chlorophyll-*a* was determined according to the Dutch standard methods NEN 6472 and 6520. Nitrite (NO_2^--N), TSS and VSS were analysed according to standard methods (APHA, 1995). Nitrate (NO_3^--N) was analysed using a Dionex ICS-100. The total nitrogen (TN) content of the biomass was determined according to NEN 6472, after digestion at ± 300ºC of the dried biomass, using salicylic acid and a H_2SO_4– selenium mixture (100 mL concentrated H_2SO_4, with 0.35 g selenium and 7.2 g of salicylic acid). The alkalinity was determined by titration with 0.020 N HCl using a methyl red indicator.

4.2.4 Biomass productivity, nitrogen and oxygen mass balance equations

The biomass productivity considered the biomass leaving the reactors (waste and in the effluent), as it would be the biomass used for further uses or production of bio-products. It was calculated using the solids wasted from the reactor to control the SRT and the solids in the effluent *(4.2)*:

$$Biomass\ productivity\ (g\ TSS\ d^{-1}) = X_R\ Q_W + X_S\ Q_S \tag{4.2}$$

The estimation of the surface biomass productivity (g TSS m^{-2} d^{-1}) was carried out considering the total biomass productivity divided by the illuminated area of the reactor. Since the reactor had a circular shape, and the light was applied around it, the illuminated area of the reactor was defined using the equation of the circumference (2Πr, where r is the radius of the circumference) and the height of the reactor.

The nitrogen mass balance (See Appendix B.1) was calculated to assess the potential nitrogen removal mechanisms. The data collected in the detailed cycles of each period was used for this purpose. The pH was controlled between 7.5 - 8.0, thus ammonium volatilization was ruled out as a significant removal mechanism (Escudero et al., 2014; González-Fernández et al., 2011b). In addition, it was assumed when ammonium was present neither nitrate nor nitrite was consumed by the microalgae. Therefore, the main nitrogen removal mechanisms assumed to have taken place were nitrification/denitrification, algal uptake and nitrogen requirements for OHO growth. Previous equations proposed to calculate the ammonium removed by partial and full nitrification (Liu and Wang, 2012) were applied in this study. The N-requirements for OHO were calculated based on

Ekama and Wentzel (2008b) and the algal uptake was calculated as described in previous reports (Mara, 2004). A complete description of the calculation steps is presented in Appendix B.1. The characterization of the biomass was calculated using the information of the N-removed through nitrification/denitrification, algal uptake and acetate oxidation by the OHOs. The biological reactions (equations) and the detailed calculations are presented in Appendix B.2.

4.2.5 Total specific and volumetric ammonium removal rate

For the different cycles studied, the total volumetric ammonium removal rates (r_{Am_T}) were calculated with Aquasim® (Reichert, 1994) using a previously proposed model (Arashiro et al., 2016) (Appendix B.3). For this purpose and in accordance to the operating and environmental conditions of each cycle in each of the periods, certain operational parameters, such as light and volume of the reactor, were adjusted within the model as well as the fractions of the biomass concentrations of algae, AOB, NOB and OHO. The model was fitted to the measured data, and the average ammonium consumption rates were estimated considering the activities of algae, nitrifiers and OHO and the results of the model. Based on the estimated ammonium removal rates, the volumetric ammonium removal rates were calculated for each period *(4.3)*:

$$r_{Am\_T} = \frac{S_{NH4\_t(0)}}{t(t) - t(0)} \tag{4.3}$$

Where:

r_{Am_T}: Total volumetric ammonium removal rate (mg NH_4^+-N L^{-1} h^{-1}).

$S_{NH4\_t(0)}$: Initial ammonium concentration (mg NH_4^+-N L^{-1})

$t(t)$: Time at which the ammonium concentration has reached zero or drops below detection limits (h)

$t(0)$: Initial time (h)

The total specific ammonium removal rate (k_{Am_T}) for the total biomass was calculated using r_{Am_T}, and the total VSS concentration using Eq. *(4.4)*:

$$k_{Am\_T} = \frac{r_{Am\_T}}{VSS} \tag{4.4}$$

Where:

k_{Am_T}: Total specific ammonium removal rate (mg NH_4^+-N mg VSS^{-1} d^{-1})

VSS: Volatile suspended solids concentrations (g VSS L^{-1}).

4.2.6 Oxygen mass balance

In order to calculate the oxygen production by algae, an oxygen mass balance was performed (Appendix B.4) over the light periods. The data used was the oxygen concentration for each of the detailed cycles, recorded every 5 minutes. In addition, the calculations of the oxygen required for ammonium oxidation by nitrifiers, acetate oxidation by heterotrophic bacteria as well as the algal and bacterial respiration were included in the balance. The oxygen transfer within the reactor was also taken into consideration using the calculated oxygen transfer coefficient. The oxygen produced by algae was calculated with Eq. B.4.3 and the endogenous respiration by OHO using the equations defined by Ekama and Wentzel (2008a). The algal respiration for the algal biomass was calculated taking into account the dark zone of the reactor for the different SRTs and applying the endogenous respiration coefficient of 0.1 d^{-1} (Zambrano et al., 2016). The oxygen consumed for nitrate production and acetate oxidation by aerobic heterotrophic bacteria were calculated using the stoichiometric expressions used for the nitrogen balance.

4.3 Results and Discussion

4.3.1 Biomass concentration and chlorophyll-*a*

The solids concentration in the photobioreactor (Figure 4.2) decreased within the first 30 days after inoculation due to the waste of biomass and the loss of biomass through the effluent. After this period, the biomass reached a maximum value of 4.2 g TSS L^{-1}. Most of the algal biomass was composed of Chlorella sp. as observed through microscopic observations. The SRT of periods 1 and 2A was similar (48 and 52 d, respectively), which may explain why, in spite of the different operational conditions, the average solids concentration was not significantly different between them ($p>0.05$). During period 2B, when the SRT was decreased to 26 days, the TSS concentration decreased to an average of 1.1 ($\pm$ 0.4) g TSS L^{-1}, which is significantly lower than in periods 1 and 2A. During period 2C, the SRT was further reduced to 17 days, but the average solids concentration (1.2 $\pm$ 0.4 g TSS L^{-1}) was not different from that of period 2B. Overall, the control of the SRT in the microalgal-bacterial consortia had an impact on the solids concentration present in the reactor.

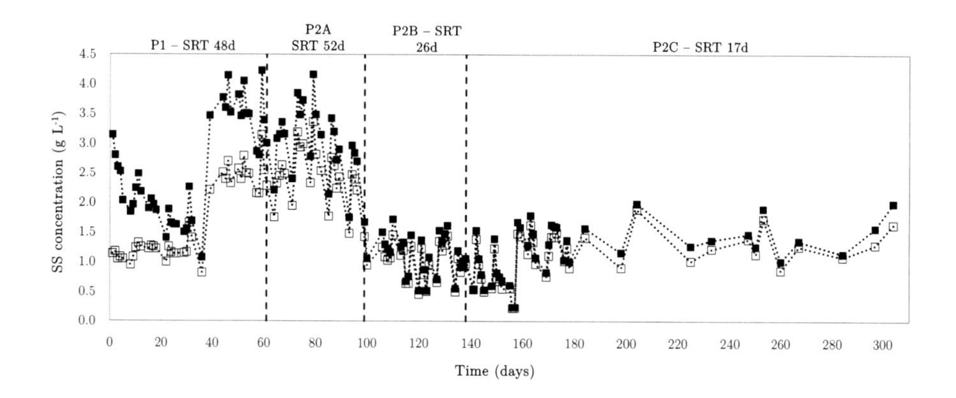

Figure 4.2. Suspended solids concentration in the sequencing-batch photobioreactor during the 4 periods. Total suspended solids concentration (TSS) (··■··), volatile suspended solids (··□··).

The highest biomass productivity (Table 4.1) was observed in period 2C, corresponding to a biomass productivity per surface area of 3.3 (± 1.2) g TSS m^{-2} d^{-1}. The higher value in period 2C was related to the higher amount of biomass wasted to ensure the SRT of 17 days. For the remaining periods, the biomass productivity was not different among them and the values were around 1.7 g TSS m^{-2} d^{-1}. These values are low in comparison with other laboratory studies on microalgal-bacterial consortia. For instance, Su et al., (2011) obtained 10.9 TSS m^{-2} d^{-1} under semi-batch operation treating municipal wastewater with an ammonium concentration between 14 and 19 mg NH_4^+-N L^{-1}. Halfhide et al. (2015) reported a biomass productivity of 2.5 g TSS m^{-2} d^{-1} in a semi-continuous system with a cell residence time of 7 days treating anaerobically digested municipal sludge centrate (220 mg NH_4^+-N L^{-1}). One of the reasons may be related to the SRT, which defines the biomass harvested from the reactor and thus the solids concentration. In addition, ammonium availability could be another explanation because the low ammonium concentration can limit the biomass production, due to the competition between nitrifiers and algae for nitrogen (Risgaard-Petersen et al., 2004).

*Table 4.1. TSS, VSS, Chl-*a *concentration, Chl-*a *content in the biomass, and biomass productivity in the 4 periods. P: Periods; BP: Biomass productivity (*Values reported for the biomass inside the photobioreactor).*

P	*SRT (d)*	*TSS* (g L^{-1})*	*VSS* (g L^{-1})*	*Chl-a* (mg L^{-1})*	*Chl-a content in biomass* (g Chl-a g VSS^{-1})*	*BP (g m^{-2} d^{-1})*	*VSS/TSS**
1	48	2.6 ± 0.9	1.7 ± 0.7	19.6 ± 10.4	0.011 ± 0.003	1.9 ± 0.6	0.65 ± 0.11
2A	52	2.9 ± 0.8	2.6 ± 0.3	28.1 ± 8.1	0.012 ± 0.002	1.7 ± 0.5	0.82 ± 0.03
2B	26	1.1 ± 0.4	1.1 ± 0.4	7.9 ± 3.0	0.008 ± 0.001	1.8 ± 0.6	0.90 ± 0.03
2C	17	1.2 ± 0.5	0.9 ± 0.2	5.1 ± 2.1	0.005 ± 0.004	3.3 ± 1.2	0.90 ± 0.05

Analysing the VSS fraction from the TSS of the biomass (Table 4.1), the fraction of inorganic solids was lower during periods 2B and 2C than in periods 1 and 2A. Therefore, an apparently higher active biomass fraction was present in the last two periods, comprising more than 85% of the VSS with regard to the TSS. This fraction is composed of biodegradable and un-biodegradable (endogenous residue) biomass (Ekama and Wentzel, 2008b). At lower SRT, the endogenous residue decreases, and correspondingly the endogenous respiration, which can have an effect on the decrease in oxygen consumed by the aerobic processes.

The average chlorophyll-*a* concentration is an indicator of the algal biomass present in a photobioreactor in relation with the total VSS. In periods 1 and 2A, the Chl-

a concentration was significantly higher than in periods 2B and 2C (Table 4.1). The highest average Chl-*a* concentration was 28.1(± 8.1) mg L^{-1} in period 2A. Based on the average values, the reduction in Chl-*a* concentration during periods 2B and 2C was a result of the SRT reduction: a higher volume of biomass was wasted to decrease the SRT, and most of the biomass was composed of algae. Thus, the Chlorophyll-*a* content in the biomass did not remain constant during the 4 periods, but instead showed the same trend as the solids concentrations. The Chl-*a* content ranged from 0.5 to 1.1%, similar to the values of 1.0 -1.5% reported in previous studies (Karya et al., 2013). However, higher values of up to 2.4%, when treating an effluent from anaerobically digested swine waste that contained 297 (± 29) mg NH_4^+-N L^{-1}, can also be found in literature (Wang et al., 2015).

4.3.2 Nitrogen and ammonium removal efficiencies and rates

Nitrogen and ammonium removal efficiency

During the 4 operational periods, the ammonium removal efficiency exceeded 94% (Table 4.2) and the ammonium effluent concentrations dropped below the detection limit (Figure 4.3). Furthermore, the total nitrogen removal efficiencies were above 70% during all the experimental periods (Table 4.2). Full nitrogen removal was not achieved, mainly because of the nitrate concentrations present in the effluent. Nitrate was not removed during part of period 1 and during most of period 2A. Nitrite concentrations were, with the exception of one day (day 165), below 0.2 mg NO_2^--N L^{-1} (Figure 4.3). Overall, the ammonium removal was successful during the 4 periods (310 days) without any external oxygen addition and despite the different operational schemes.

Table 4.2. Ammonium removal efficiencies and rates for the 4 operational periods

P	*SRT (days)*	*ARE (%)*	*NRE (%)*	r_{Am_T} (mg NH_4^+-N L^{-1} h^{-1})	k_{Am_T} (mgNH_4^+-N mgVSS^{-1} d^{-1})	r_{N_T} (mg NH_4^+-N L^{-1} h^{-1})	k_{N_T} (mgNH_4^+-N mgVSS^{-1} d^{-1})
1	48	94.3±23.3	87.2±25.7	2.94	0.061±0.005	1.9±0.3	0.029±0.013
2A	52	100±0.0	73.9±24.7	1.85	0.017±0.002	1.4±0.5	0.017±0.010
2B	26	99.9±0.4	89.5±1.5	1.71	0.042±0.022	1.7±0.1	0.048±0.019
2C	17	96.6±15.1	87.0±16.7	2.12	0.063±0.009	1.6±0.3	0.049±0.039

P: periods; ARE: ammonium removal efficiency; NRE: total nitrogen removal efficiency; r_{Am_T}: Volumetric ammonium removal rate; k_{Am_T}: specific ammonium removal rate; r_{N_T}: Volumetric nitrogen removal rate; k_{N_T}: Specific nitrogen removal rate

During period 1 and 2A, the denitrification process was limited and the concentrations in the effluent reached up to 11.5 mg NO_3^--N L^{-1} in period 1 and 19.8 mg NO_3^--N L^{-1} in period 2A. During these periods, it appears that the concentration of organic carbon was limiting the denitrification activity. Furthermore, the denitrification process was further limited at longer SRTs when reaching the highest concentrations of solids (Figure 4.2). Most of the biomass was composed of algae, more specifically *Chlorella sp.* (microscopic observations, data non shown). *Chlorella sp.* are known to grow mixotrophically (Perez-Garcia et al., 2011; Wu et al., 2014), especially under dark conditions. Therefore, possibly, a competition for organic carbon between algae and OHO occurred in the dark phases of periods 1 and 2A. For instance, in period 1 after 40 days, the denitrification process ceased, coinciding with the higher biomass concentration (Figure 4.2). Once the SRT decreased to 26 days, consequently decreasing the solid and algal concentrations, the denitrification process resumed and total nitrogen removal was achieved in periods 2B and 2C. These observations support the hypothesis that algae and OHO may compete for organic carbon sources at longer SRT, which can lead to higher concentrations of solids.

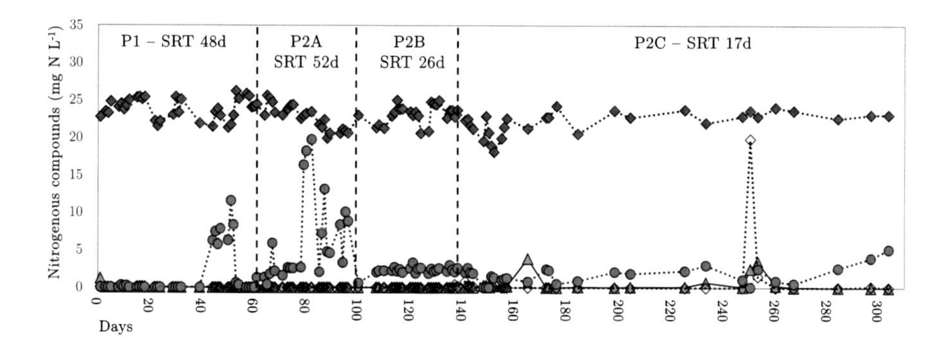

Figure 4.3. Daily nitrogenous concentrations in the reactor along the experimental periods. Influent NH_4^+-N (—◆—), effluent NH_4^+-N (—◇—), effluent NO_3^--N (—●—), and effluent NO_2^--N (—▲—).

Figure 4.4 presents an example of the evolution of the N-compounds and oxygen concentration trends during one cycle in period 1 (day 47) and one cycle in period 2B (day 117). The trends of the N-compounds and oxygen concentration for period 2B were similar in periods 2A and 2C. In this particular comparison among these two cycles, it can be observed that the conversion of ammonium to nitrate was higher in period 1 than 2B, this is due the different feeding schemes. During periods 2A, 2B and 2C, the ammonium and organic carbon sources were supplied simultaneously (at the beginning of the dark phase), hence part of the organic carbon that was not used for denitrification or not consumed by *Chlorella sp.* in period 2A was oxidized aerobically. This can lead to a competition for oxygen between AOBs and OHO, resulting in lower ammonium to nitrate conversion rates. Certainly, the feeding regime proposed in period 1 is not realistic since ammonium and organic carbon within the wastewater cannot be selectively separated. Therefore, the operational conditions were modified in the last 3 periods (2A, 2B and 2C) to create more realistic operating conditions.

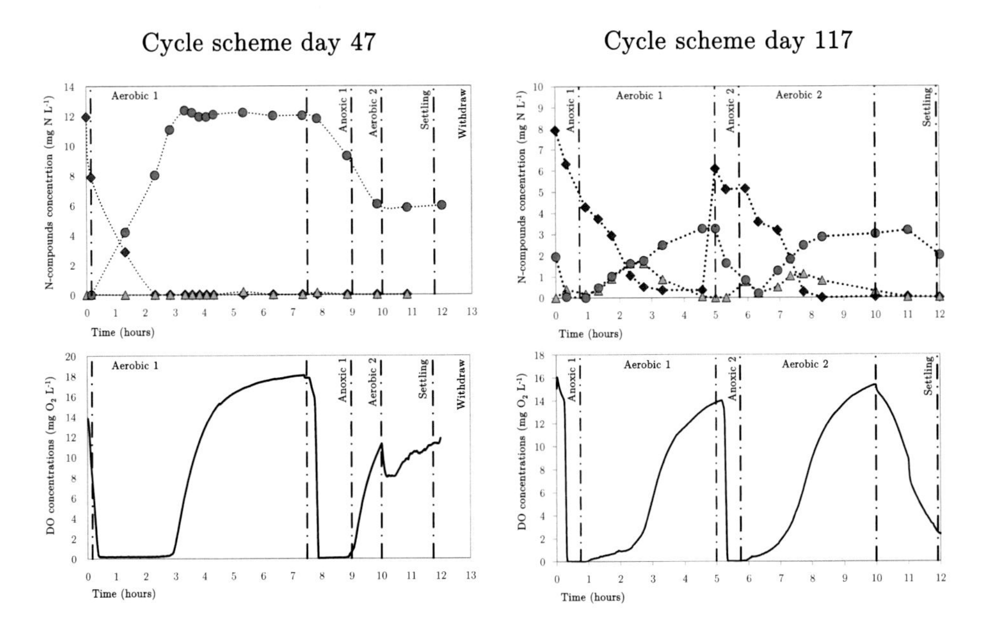

Figure 4.4. Variation of nitrogen compounds and dissolved oxygen during a SBR cycle scheme for day 47 (Period 1) and day 117 (Period 2B). The trends of the N-compound and oxygen concentrations during period 2B were similar to periods 2A and 2C. Anoxic refers to the dark periods and aerobic to the light periods. NH_4^+*-N (—◆—),* NO_3^-*-N (—●—), and* NO_2^-*-N (—▲—).*

Ammonium and total nitrogen removal rates at different SRTs

The total volumetric ammonium removal rate was calculated for the 4 periods (Table 4.2) using the algal-bacterial model in Aquasim (Reichert, 1994). The highest ammonium removal rate among the 4 periods (r_{Am_T}) was 2.94 mg NH_4^+-N

L^{-1} h^{-1} for period 1. The higher removal rate in period 1 compared with periods 2A, 2B and 2C could be a consequence of the feeding operation rather than the SRT (period 2A and 1 had similar SRT times). The separated feeding of organic carbon and ammonium favoured the nitrification in period 1, since there was no O2 competition between AOBs and OHOs with regard to the oxidation processes.

The r_{Am_T} varied among periods 2A, 2B and 2C (same operational scheme) for the different SRTs tested, but it did not show any clear trend as there is a decrease in r_{Am_T} from period 2A to 2B, but the fastest ammonium removal was observed in period 2C (2.12 mg NH_4^+-N L^{-1} h^{-1}) at an SRT of 17 days. An r_{Am_T} for a microalgal-bacterial reactor of 7.7 mg NH_4^+-N L^{-1} h^{-1} have been previously reported (Karya et al., 2013), treating synthetic wastewater at an SRT of 15 days. While values of around 4.1 mg mg NH_4^+-N L^{-1} h^{-1} at a HRT of 1 day and SRT of 2.6 days have also been observed (Chapter 3). Herein, the rates calculated for periods 2A to 2C are similar to those reported by other researchers with an ARR of 2.3 mg mg NH_4^+-N L^{-1} h^{-1} in a continuous laboratory-scale microalgal-bacterial system treating synthetic wastewater at a SRT of 15 days and HRT of 1 day (van der Steen et al., 2015). Ammonium removal rates in the range of 1.5 mg NH_4^+-N L^{-1} h^{-1} have also been reached (Molinuevo-Salces et al., 2010) in a closed reactor with a HRT of 10 days treating a diluted anaerobically digested slurry.

In contrast to the r_{Am_T}, when comparing the average values for periods 2A, 2B and 2C (performed under similar operational conditions), the highest specific ammonium removal rates increased at shorter SRTs reaching a maximum specific rate of 0.063 ($\pm$ 0.009) mg NH_4^+-N mgVSS^{-1} d^{-1} in period 2C. This value is a reflection of the dual action of nitrifiers and algae. Furthermore, this indicates that the biomass was significantly more active at shorter SRT, being able to remove more ammonium per gram of biomass. For cultures of algae and nitrifying bacteria, k_{Am_T} of 0.02 mg NH_4^+-N mgVSS^{-1} d^{-1} and 0.05 mg NH_4^+-N mgVSS^{-1} d^{-1}, respectively, have been found in literature (Vargas et al., 2016), for a total value in a microalgal-bacterial consortium of 0.07 mg NH_4^+-N mgVSS^{-1} d^{-1}. In a pilot-scale HRAP operated to cultivate algae using domestic wastewater (containing 39.7± 17.9 mg NH_4^+-N L^{-1}), the k_{Am_T} was 0.03 mg NH_4^+-N mgVSS^{-1} d^{-1} (Sutherland et al., 2014). In another study, the highest specific removal rate reached 0.05 mg TN mg TSS^{-1} d^{-1} (Posadas et al., 2013) in a pilot-scale raceway. Noteworthy, the latter study reported the k_{Am_T} rate in terms of TSS, which can underestimate the specific removal rate, as it includes the inorganic fraction of the biomass.

The k_{Am_T} of nitrifying bacteria can vary between 0.5 and 5.2 mg NH_4^+-N mgVSS^{-1} d^{-1} (Wiesmann, 1994a). The reason for this wide range of values can be due to the biomass characterization. Despite that the ammonium removal through nitrifying bacteria in a microalgal-bacterial system can account for more than 50% of the total removal (Su et al., 2011; Karya et al., 2013; Van Den Hende et al., 2014; Wang et al., 2015), the nitrifying biomass does not comprise more than 6% of the total biomass (Karya et al., 2013; Chapter 3). Other studies using microalgal-bacterial systems using synthetic wastewater have reported values of 0.1 and 1.4 mg NH_4^+-N mgVSS$_{nitrifiers}^{-1}$ d^{-1}, in which nitrification was the main ammonium removal mechanism (Karya et al., 2013; Chapter 3). Moreover, based on the results of the specific nitrogen removal rates, and just taking into account periods 2A, 2B and 2C (performed under similar conditions) (Table 4.2), it can be observed that the rates in period 2B and 2C are significantly higher than in period 2A, suggesting that shorter SRTs also tend to favour the nitrogen removal rates of microalgal-bacterial consortia. Likely, lower SRTs increased the specific ammonium removal rates due to an increase in the ammonium loading rate, which results in an increase in the ammonium oxidation rate (Pollice et al., 2002) once all the other conditions are met, and as far as the applied SRT is not shorter than the minimum required SRT for nitrification (Ekama and Wentzel, 2008b).

4.3.3 Nitrogen removal mechanisms and biomass characterization

Nitrogen removal mechanisms

The contribution of the nitrogen removal mechanisms to the total N removal were estimated in order to identify if there was a correlation among them and the SRT tested. The calculations were made only with detailed data from cycles of periods 2A, 2B and 2C, since they were performed under similar operating conditions. Based on the results of section 4.3.2, the ammonium and nitrogen removal efficiencies did not differ among the different SRT tested. Also, the amount of ammonium removed during the three periods was not significantly different ($p > 0.05$) and remained around 13.9 ($\pm$ 1.4) mgNH_4^+-N d^{-1} (Table 4.3). The ammonium removal mechanisms identified were nitrification/denitrification and nitrogen consumption due to the growth requirements of OHOs and algal uptake (Table 4.3).

Nitrification/denitrification was the main removal mechanism (44 - 74%), the remaining ammonium was removed by algae assimilation (11 - 38%) and a small portion was used for bacterial growth (OHOs and nitrifiers). Other studies have also reported the successful removal of ammonium and TN up to negligible values at laboratory and pilot scale, when using microalgal-bacterial consortia without any supply of external air or oxygen. This shows the ability of the system to benefit from the symbiosis of algae and bacteria (Arashiro et al., 2016; García et al., 2017; Liang et al., 2013; Chapter 3; Solimeno et al., 2017; van der Steen et al., 2015; Wang et al., 2015).

With further data analysis, the maximum nitrate formation rate occurred in period 2C (of 10.3 ± 2.8 $mgNO_3^-$-N d^{-1}), which corresponded to the shortest SRT tested, and the highest denitrification rate was found in periods 2B and 2C. The lowest nitrogen consumption for biomass synthesis of heterotrophic bacteria was observed in period 2A (at a SRT of 52 days), which was somehow expected since the nitrogen requirements for biomass growth decreases as the sludge age increases (Ekama and Wentzel, 2008b). The uptake of ammonium by algae was higher at longer SRTs, like in period 2A and 2B, and was not significantly different among them ($p>0.05$), while the lowest (1.5±0.9 $mgNH_4^+$-N d^{-1}) was observed in period 2C at the short SRT of 17 days. Despite that the net amount of ammonium removed was not different between the three periods, the main removal mechanisms were different, and the nitrification rate was higher at the shortest retention time tested. At longer SRTs, the main removal mechanism was through algae assimilation followed by nitrification, while the opposite took place at shorter SRTs where nitrification was the main removal mechanism. The differences might be attributed to the oxygen conditions during the different SRTs, which is an indirect result of the different solid concentrations and biomass characteristics.

Biomass characterization

The biomass was mainly composed of algae and heterotrophic bacteria (Figure 4.5). Based on the removal mechanisms (Table 4.3), the theoretical biomass characterization shows that the total VSS was composed of 40 to 70% by algae. The OHOs comprised between 25 and 50%, and their highest fraction was estimated in period 2C. Two to 7% of the total VSS was composed of nitrifiers, and similar to OHO their highest fraction was observed in period 2C, when the highest ammonium to nitrate conversion was observed. The lowest algae content was observed in period 2C, this value is mainly due to the lower algal uptake measured in this period (Table 4.3), which is in line with the chlorophyll-*a* concentrations. This suggests that shorter SRTs also had an effect on the bacterial composition. When comparing these results to those from other studies, in a HRAP treating municipal wastewater at an HRT of 4 days, the biomass was composed of 56-78% by algal biomass (Solimeno et al., 2017), while 30-20% by bacteria (including OHO and nitrifiers). In another study, the biomass composition contained 67% algae, 16% OHO and 17% nitrifiers in a lab-scale photobioreactor treating diluted centrate from an anaerobic digester used to process swine manure (Arashiro et al., 2016). The differences are attributed to the different operational conditions and cultivation medium. Nevertheless and despite those differences, algae dominated the microalgal-bacterial biomass in all those studies.

The biomass fractionation was calculated based on the VSS concentration (Appendix B.3). Therefore, it must be taken into account that an endogenous residue (e.g. of non active biomass within the VSS) is accounted for within this value (Ekama and Wentzel, 2008b). This endogenous residue depends on the SRT: it is higher at longer SRT for activated sludge (Ekama and Wentzel, 2008b). In the case of algae, the high solids concentration can increase the dark zones within the reactor causing light attenuation, and in parallel increase the endogenous respiration of algae. Consequently, the control of the VSS through the SRT can affect the oxygen consumption and the biomass composition.

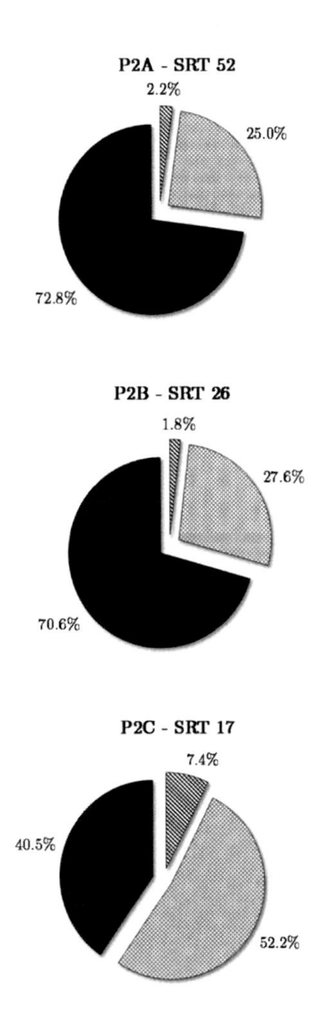

Figure 4.5. Biomass composition at the different SRTs tested. VSS nitrifiers (▧), VSS OHO (▩), VSS Algae (■).

Table 4.3. Nitrogen removed by the different mechanisms for the different operational periods.

Period	*SRT (days)*	*Total NH_4^+-N removed (mg NH_4^+-N d^{-1})*	*Total NO_3^--N formed (mg NO_3^--N d^{-1})*	*NH_4^+-N removed by nitrification (mg NH_4^+-N d^{-1})*	*Total NO_3^--N denitrified (mg NO_3^--N d^{-1})*	*$OHO-N$ requirement (mg NH_4^+-N d^{-1})*	*NH_4^+-N removed by algae (mg NH_4^+-N d^{-1})*
2A	52	13.2±1.1	6.5±0.4	6.7±0.4	2.7±0.0	1.7±0.0	4.9±0.8
2B	26	14.0±0.7	6.3±0.3	6.3±0.3	4.9±0.3	2.3±0.6	5.4±0.4
2C	17	14.2±2.2	10.3±2.8	10.7±2.9	4.6±1.5	2.0±0.4	1.6±0.9

4.3.4 Oxygen production in a microalgal-bacterial photobioreactor under different SRTs

The oxygen production by algae and the oxygen consumption by the different aerobic processes was calculated (Table 4.4 and Table 4.5). The highest production of oxygen by algae was 0.33 kg O2 m^{-3} d^{-1} during period 2A, while for periods 2B and 2C, there were no significant differences ($p>0.05$). The total O_2 produced by algae was sufficient to sustain the total oxygen consumption within the photobioreactor (nitrification, algal and bacterial respiration and COD oxidation).

A higher oxygen production rate of up to 0.46 kg O_2 m^{-3} d^{-1} for a microalgal-bacterial system using synthetic domestic wastewater has been reported previously (Karya et al., 2013). The different oxygen production rates cannot be directly explained based on nitrogen uptake (as a means to assess the algal activity) since the ammonium uptakes are similar in both studies (4.7 ± 2.7 mg NH_4^+-N L^{-1} (Karya et al., 2013) *versus* 4.9 ± 0.8 mg NH_4^+-N L^{-1} during period 2A). Possibly and in spite of the similar ammonium uptake, the difference in O_2 production could be due to light limitation. Actually, the higher biomass concentrations reported in period 2A, that likely induced light attenuation, are higher than those observed elsewhere (Karya et al., 2013). Furthermore and interestingly, in this study the oxygen production did not cease when the ammonium was consumed (Figure 4.4). Moreover, the microalgal-bacterial biomass in this research did not take up nitrate during the conduction of ex-situ batch tests (data not shown) that could be used as an alternative source of nitrogen. As such, there may be other additional or alternative N sources for algae to grow and perform photosynthesis. Nevertheless, algae are capable of storing N under N-stress conditions (Lavín and Lourenço, 2005).

Probably, this mechanism took place in this study, as observed previously (Wágner et al., 2016), when ammonium was exhausted. Further studies are required to assess the potential mechanism of oxygen generation by algae via the potential use of intracellularly stored nitrogen compounds.

Table 4.4. Total oxygen produced and consumed in the SBR. P: Periods

P	*SRT*	*O_2 produced by algae (kg O_2 m^{-3} d^{-1})*	*O_2 transfered (kg O_2 m^{-3} d^{-1})*	*O_2 consumed (kg O_2 m^{-3} d^{-1}1)*
2A	52	0.33±0.04	0.03±0.00	0.32±0.05
2B	26	0.25±0.04	0.02±0.01	0.19±0.03
2C	17	0.21±0.02	0.03±0.02	0.20±0.02

Table 4.5. Oxygen consumption by the different aerobic and endogenous respiration processes. P: Periods

P	*SRT*	*O_2 nitrification (kg O_2 m^{-3} d^{-1})*	*O_2 COD oxidation (kg O_2 m^{-3} d^{-1})*	*O_2 algae respiration (kg O_2 m^{-3} d^{-1})*	*O_2 OHO respiration (kg O_2 m^{-3} d^{-1})*
2A	52	0.06±0.01	0.06±0.04	0.131±0.015	0.07±0.01
2B	26	0.06±0.00	0.02±0.01	0.025±0.009	0.08±0.02
2C	17	0.11±0.03	0.03±0.02	0.005±0.002	0.05±0.01

Based on an oxygen mass balance, an oxygen production of 0.19 kg O_2 m^{-3} d^{-1} for a microalgal-bacterial system treating real anaerobic digested swine waste centrate at a SRT of 8 days and HRT of 4 days was estimated previously (Wang et al., 2015). That lower O_2 production compared with the values found in this research (Table 4.4) might be caused by light limitation. In that study (Wang et al., 2015), the oxygen concentration did not exceed 0.5 mg O_2 L^{-1}, which the author attributed to the turbidity of the swine waste centrate.

Herein, the consumption of oxygen was higher at the SRT of 52 than at the SRT of 26 or 17 days. However, different aerobic processes were involved in the oxygen consumption at an either long or short SRT. During the SRT of 52 days (period 2A), COD oxidation accounted for 19% of the oxygen consumption, and it was higher than in periods 2B and 2C. Also, at a SRT of 52 days, the denitrification process deteriorated (Figure 4.3). Therefore, part of the COD supplied was not consumed by denitrifiers in the dark period, but mostly oxidised in the aerobic phase. Despite these issues, the COD removal efficiencies were 89%, 84% and 88% in the experimental periods 2A, 2B and 2C, respectively. It appears that the respiration of OHO was not significantly different among these periods ($p>0.05$) meanwhile the algal respiration decreased at shorter SRTs. It was higher in period 2A, which accounted for 41% of the total oxygen consumption, while during period 2C (SRT 17 days) algae respiration represented only 2% of the total O_2 consumption (Table 4.5).

During period 2C, nitrification had the highest oxygen consumption, accounting for 54% of the total O_2 consumed (Table 4.5). This is in line with the findings of the nitrogen balance since in period 2C the highest formation of nitrate occurred. The higher nitrification rates observed in period 2C could be associated with a higher availability of oxygen in comparison with period 2A (Table 4.5). As the reactor is operated as a sequencing batch reactor, oxygen limitation could occur as soon as the medium was fed. During period 2A, there was a higher algal respiration and COD oxidation; therefore, during the first hours of the cycle, oxygen was limited at longer SRT. Similar findings have been reported in literature (Arashiro et al., 2016) who compared two photobioreactors at different SRT (7 and 11 days). In that study, at shorter SRT the NH_4^+ conversion to nitrite was higher than at longer SRT, attributed to the higher availability of oxygen at the shorter SRT. Therefore, longer SRTs can cause oxygen limitation, hindering the oxidation of ammonium by AOB and the oxidation of organic carbon. Ultimately, this will decrease the efficiency of the system.

With regard to algae respiration, it must be taken into account that algae respire at higher rates under dark conditions. The dark zones were estimated using the Lambert-Beert equation and the total solids concentration for each period. The dark zones fractions calculated were 0.78, 0.36, and 0.17 for periods 2A, 2B and 2C, respectively. Since the respiration rate is assumed to remain constant at a rate of 0.1 d^{-1} (Zambrano et al., 2016), the combination of the higher algal concentration and the dark zones might have caused an increase in algal respiration at longer SRTs. However, in spite of the light limitation at longer SRTs, the oxygen production was not considerably different among the SRTs, and the highest O_2 production took place in period 2A. This can be related to the higher ammonium uptake observed in period 2A, compared with periods 2B and 2C. Nevertheless, microalgae have the capacity to adapt to either light limited environments or higher light irradiance. In a pilot-scale HRAP, in which the HRT was equal to the SRT (not decoupled), a decrease in light availability at longer HRT (and therefore with higher biomass concentrations) was also observed (Sutherland et al., 2015). However, the photosynthetic efficiency (rate of photosynthesis) increased at longer HRT and higher depths. Under a higher light intensity, algae regulate the light absorption by decreasing the chlorophyll content per cell (Bonente et al., 2012), while under low light conditions algae increase the synthesis of photosynthetic systems within the cell to convert more light (Falkowski and Raven, 2013). Nonetheless, the light attenuation by the biomass in the reactor has a direct implication for the O_2 production. Therefore, in order to supply sufficient oxygen for the aerobic process without causing O_2 saturation, or oxygen limited conditions, the control of the biomass in the reactor should be ensured by controlling the SRT.

4.3.5 Effects of SRT on the light penetration, ammonium removal mechanisms and oxygen production

This study shows that high ammonium removal rates can be achieved in microalgal-bacterial systems at higher volumetric and specific rates through nitrification/denitrification when operated at shorter SRTs. The SRT controls the solids concentration, which affects the light conditions inside the reactor, as well as the respiration rates. This will concomitantly affect the oxygen production and availability. Nonetheless, it must be taken into account that the SRT should not fall below the minimum required SRT for nitrification (a SRT at which the dilution of biomass is higher than the doubling time of the nitrifiers), otherwise this will lead to the wash out of nitrifying organisms from the system (Ekama and Wentzel, 2008a). The minimum SRT calculated for this system is 2.9 days, and at lower SRT the nitrification rates will start to decrease exponentially (Ekama and Wentzel, 2008a). When testing different SRT between 5 and 2 days in a sequencing batch nitrifier reactor, partial nitrification was observed at an SRT below 2 days, while for SRTs of 3, 4 and 5 days full nitrification was achieved (Munz et al., 2011). In activated sludge systems, sludge ages of 10 to 25 days are recommended for biological nutrient removal, a value usually 5 to 8 days longer than that applied in systems that achieve only COD removal (Ekama and Wentzel, 2008b).

Thus, to achieve nitrogen removal via ammonium oxidation and further denitrification, microalgal-bacterial systems can be operated at shorter SRT, without have higher HRT. Shorter SRT improve light conditions inside the reactor by ensuring low biomass concentration, while avoiding biomass wash out, which ultimately promote nitrification over algal uptake, as long as other conditions are met (e.g. C/N/P ratio, pH, and temperature). Therefore, uncoupling of the SRT from the HRT, allows to select SRTs that promote higher ammonium removal rates by optimizing the O_2 production, decrease of the light attenuation, and higher biomass productivities while requiring less area.

The SRT is a key operational parameter that allows to control the efficiency and growth of the microalgal-bacterial systems (at constant conditions), specially under different environments in which the incident light cannot be modified. During seasonal variations, when light is limited or there is a lower temperature, the biomass concentration in the reactor can be adjusted to keep higher removal efficiencies. Based on the results and literature data, the optimal SRT and HRT for the design of a microalgal-bacterial system treating municipal wastewater is between a SRT of 10 to 17 days with a HRT of 1 day. However, it must be noted that these conclusions were made using laboratory-scale experiments, therefore these results must be tested at pilot and large scale. Furthermore, they may vary depending on the environmental conditions and influent characteristics of the wastewater of interest.

4.4 CONCLUSIONS

Ammonium and total nitrogen were successfully removed at a SRT of 52, 26 and 17 days. The SRTs had an impact on the removal mechanisms. Nitrification was the main removal mechanism, achieving higher nitrate formation at shorter SRTs. Ammonium removal through nitrification increased from 50.5% (SRT 52 d) to 74.3% (SRT 17 d). The highest ammonium removal rate was 2.12 mg mgNH_4^+-N L^{-1} h^{-1} with a specific removal rate of 0.063 gNH_4^+-N gVSS^{-1} h^{-1} at a SRT of 17 days. Shorter SRT improve light conditions inside the reactor by reducing the solids concentration, which has a direct positive effect on the oxygen production and consumption. This study suggests that higher ammonium removal rates for microalgal-bacterial systems operated as sequential batch reactor can be achieved at shorter SRTs and HRTs.

5

MODELLING OF NITROGEN REMOVAL USING A MICROALGAL-BACTERIAL CONSORTIUM

This chapter is based on: Arashiro, L.T., Rada-Ariza, A.M., Wang, M., Steen, P. van der, Ergas, S.J., 2016. Modelling shortcut nitrogen removal from wastewater using an algal-bacterial consortium. Water Science and Technology 75, 782–792. https://doi.org/10.2166/wst.2016.561

Abstract

The treatment of high ammonium strength wastewater was achieved using an algal-bacterial consortium in two photo-sequencing batch reactors (PSBRs). The nitrogen removal mechanisms were nitritation/denitritation, in this process, algae provide oxygen for nitritation during the light period, while denitritation takes place during the dark (anoxic) period, reducing overall energy and chemical requirements. The two PSBRs were operated at different solids retention times (SRTs) and the ammonium concentration in the wastewater fed was 264 mg NH_4^+-N L^{-1}, with a 12 hour on/12 hour off light cycle. The average surface light intensity was 84 μmol $m^{-2}s^{-1}$. The total inorganic nitrogen removal efficiencies for the two PSBRs was ~95%, and the biomass settleability was measured as SVI (53-58 mLg^{-1}). Higher biomass density was observed at higher SRT, resulting in greater light attenuation and less oxygen production. A mathematical model was developed to describe the algal-bacterial interactions using the Activated Sludge Model No.3 (ASM3) as base, and including two algal processes. The results of the model predicted the experimental data closely. One of the most sensitive parameters was found to be the maximum growth rate of algae ($\mu_{max,P}$).

5.1 INTRODUCTION

Anaerobic digestion (AD) of domestic, industrial and agricultural wastes stabilizes organic matter and produces biogas that can be used as an energy source. However, effluents from AD contain high NH_4^+-N concentrations, which can induce eutrophication in natural waters. The conventional biological nitrogen removal pathway for such effluents is the combination of nitrification and denitrification. Innovative shortcut nitrogen removal processes (i.e. nitritation-denitritation) have been developed over the past decade that save up to 25% of energy for aeration and 40% of carbon source requirements compared with conventional nitrification-denitrification processes (Wiesmann et al., 2006). Aeration costs could be further reduced by using algae photosynthesis for oxygen supply. Studies have shown that wastewater treatment systems containing algal-bacterial consortia may provide additional energy savings and higher nutrient removal efficiency, when compared to systems that rely only on either algal or bacterial processes (Liang et al., 2013). This algal-bacterial symbiosis can be applied in photobioreactors to reduce the concentrations of nutrients while reducing the electrical energy demands from aeration in wastewater treatment processes (Kouzuma and Watanabe, 2015). In these reactors, the photosynthetic activity of microalgae provides oxygen needed for organic matter oxidation and nitrification during the light periods. Denitrification or denitritation processes take place primarily during the dark (anoxic) period.

The availability of light inside the photobioreactor is a major factor for microalgal photosynthesis, affecting the oxygen production process. Light availability is affected by concentrations of dissolved organic compounds and total suspended solids (TSS), which are related to the photobioreactor operating conditions, particularly the solids retention time (SRT). A SRT of 15 days was shown to result in complete nitrification without mechanical aeration in a study using a consortium of algae and nitrifiers to treat synthetic wastewater (50 mg NH_4^+-N L^{-1}) in photo-sequencing batch reactors (PSBRs) (Karya et al., 2013). Wang et al. (2015) treated centrate from anaerobically digested swine manure with higher ammonium concentration (300 mg NH_4^+-N L^{-1}) and also achieved complete ammonia removal via nitritation-denitritation in PSBRs with alternating light and dark periods and SRT of 8 days.

Although these authors and others (de Godos et al., 2014) recently studied algal-bacterial symbiosis for wastewater treatment there is still a lack of research on modelling the performance of algal-bacterial systems. Models are needed to predict, for example, growth of both microorganisms, efficiency of nutrient removal from wastewater during different seasons and in different geographic regions or the effect of system design and operational parameters on overall system performance. One of the latest biological process models for use in wastewater treatment is the Activated Sludge Model no. 3 (ASM3), which better describes the decay processes compared to ASM1 and includes cell internal storage compounds (Henze et al. 2000). However, a disadvantage of ASM3 is the representation of nitrification and denitrification as single-step processes. Thus, the activities of the ammonium oxidizing bacteria and archaea (AOB and AOA) and nitrite oxidizing bacteria (NOB) are not properly distinguished. In order to be able to describe shortcut nitrogen removal, nitrite dynamics in wastewater treatment systems should be modelled. Some researchers therefore have proposed new versions of ASM3, extended to two-step nitrification and two-step denitrification, i.e. with nitrite as an intermediate (Iacopozzi et al., 2007; Kaelin et al., 2009). Models for bacterial growth could be combined with models for algal growth. Several researchers have suggested mathematical models to describe algal photosynthesis and growth kinetics, which can be expressed as a function of light conditions (Halfhide et al.,

2015; Martinez Sancho et al., 1991), temperature and pH (Costache et al., 2013), and inorganic carbon, inorganic nitrogen and inorganic phosphorus (Decostere et al., 2016).

This chapter reports on experimental PSBR studies and the development and calibration of a mathematical model that represents the performance of the algal-bacterial PSBR under varying operating conditions. The model describes how light availability is affected by dissolved and suspended matter concentrations in the PSBR and how light attenuation influences oxygen production and nitrogen removal.

5.2 Materials and Methods

5.2.1 Experimental

Photo-sequencing batch reactor

The design and operation of the bench-scale PSBRs used in this study have been described elsewhere (Wang et al., 2015). Briefly, two cylindrical glass reactors (2L volume, 16 cm diameter, 10 cm height) were inoculated with a mixed microbial consortium, which contained nitrifying and heterotrophic bacteria derived from a wastewater mixed liquor seed and wild strain algae - mainly *Chlorella* spp. The PSBRs were fed with centrate from a pilot-scale mesophilic anaerobic digester that was used to process swine manure, which was collected from Twenty Four Rivers Farm (Plant City, FL) on a weekly basis, mixed with urea and local groundwater and fed to the digester three times per week. Urea was added to make up for the loss of urine due to the farm operation. The swine centrate was centrifuged for 15 min at a speed of 4000 rpm, filtered with a 0.45μm membrane filter and diluted three times before being used to feed the PSBRs, with an average NH_4^+-N concentration of 264±10 mg L^{-1}. Typical characteristics of the influent are shown in Table 5.1. Despite the centrifugation and filtration the soluble fraction of the COD represented 72% of the total COD, similar values (76 %) were found by Wang et al., (2015), who used the same influent following the same pre-treatment of centrifugation and dilution.

Table 5.1. Typical characteristics of the influent (diluted swine centrate).

Variable	*Value*
pH	6.8 ± 0.1
Total COD (mg/L)	810 ± 82
Soluble COD (mg/L)	589 ± 71
NH_4^+-N (mg/L)	264 ± 10
NO_2^--N (mg/L)	4.0 ± 0.3
NO_3^--N (mg/L)	0.3 ± 0.1
TN (mg/L)	313 ± 137
Alkalinity (mg $CaCO_3$/L)	1574 ± 46
TP (mg/L)	27.5 ± 4.2
Soluble TP (mg/L)	22.9 ± 4.7

The operation of the PSBRs consisted of a 24 hour cycle (feed, react, settle, decant), of which 12h were illuminated and 12h were dark. The PSBRs were continuously stirred at 200 rpm using a magnetic mixer, except during settling and withdrawal stages at the end of the dark period. The temperature inside both PSBRs was 27 (± 3) °C. The experiment was divided into two phases (Figure 5.1); in Phase 1 no external carbon source was applied while during Phase 2 sodium acetate was added at the start of the dark period to promote denitrification, based on the stoichiometry of 2.2 g COD g^{-1} NO_2^--N removed. Between Phase 1 and Phase 2, a 4-day dark period was applied when sodium acetate was added (amount needed for full denitrification) to the PSBRs to eliminate accumulated NO_2^--N from Phase 1. No inflow or outflow were introduced during the 4-day dark period. No CO_2 was added during the operational steps since alkalinity was sufficient for the nitrification and algae growth (1574 mg $CaCO_3$ L^{-1}).

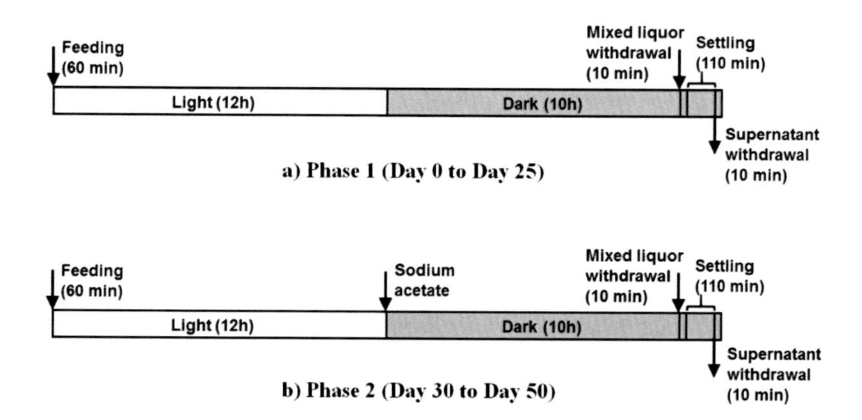

Figure 5.1. Operational steps of the PSBRs during one cycle of a) Phase 1: no sodium acetate addition and of b) Phase 2: with sodium acetate addition at the start of the dark period.

The hydraulic retention time (HRT) was maintained at 4 days in both PSBRs, but each reactor was operated at a different SRT. Reactor 1 (R1) was operated with an average SRT of 7 days and Reactor 2 (R2) of 11 days. SRTs were maintained by withdrawing a portion (R1: 250 mL, R2: 150mL) of the mixed liquor each day, just before the settling period. The SRT was calculated by equation *(5.1)*:

$$\text{SRT (d)} = \frac{TSS_R\ V_R}{TSS_R\ V_W + TSS_E\ V_E} \tag{5.1}$$

where TSS_R is the biomass concentration of the mixed liquor (mg L^{-1}); TSS_E is the biomass concentration of the effluent (mg L^{-1}); V_R is the reactor volume (L); V_W is the daily volume of wasted mixed liquor (L d^{-1}) and V_E is the daily volume of effluent (L d^{-1}).

Incident light

The PSBRs were irradiated with two banks of eight cool white fluorescent tubes (Philips Cool White-20W, 24 inches), placed on two sides of the reactors providing an average light intensity on the surface of the PSBRs of 84 ± 3 µmol m^{-2} s^{-1}. Incident light intensity was measured with a Quantum meter MQ-200 (Apogee Instruments, US) at eight different points along the reactors' wall and the light intensity considered for both PSBRs is given as the average value of these measurements.

Light attenuation

The light intensity (I) within the PSBRs cannot be merely represented by the light intensity at the surface of the PSBRs. Light attenuation causes a considerable reduction in light intensity along the depth of the reactor. The modified Beer-Lambert law was applied to describe the light intensity at a specific position from the light source as (Martinez Sancho et al., 1991):

$$I(x) = I_0 \exp(-kX_T x) \tag{5.2}$$

where I_0 is the initial light intensity (μmol m^{-2} s^{-1}), k is the extinction coefficient (m^2 g^{-1} TSS), X_T is the TSS concentration (g TSS m^{-3}) and x is the distance from the light source (m).

The light intensity was measured at nine different points along the reactor radius (every 1 cm distance from 0 cm to 8 cm), at varying distance from the light source, inside one of the PSBRs using a Quantum meter MQ-200 (Apogee Instruments, US). This procedure was repeated with seven different concentrations of mixed liquor and influent to study the influence of TSS concentration on the light availability inside the PSBR. All the dilutions were made using the influent and the first concentration (C1) corresponds to the influent without any algal-bacterial biomass. The data collected from this experiment were used to determine the extinction coefficient, *k*, in Eq. 5.2 using the MS Excel tool *Solver* (GRG nonlinear algorithm).

Analytical methods

pH and dissolved oxygen (DO) were measured with an Orion GS9156 pH and DO meter (Thermo Fisher Scientific Inc., Waltham, MA, US), respectively, and calibrated electrodes. Chlorophyll-*a* was measured using the ethanol extraction method according to NEN 6520 – Dutch Standard (NEN 2006). TSS and volatile suspended solids (VSS) were measured according to Standard Methods 2540 D (APHA 2012). The concentrations of NH_4^+, NO_2^- and NO_3^- were measured using a Metrohm Peak 850 Professional AnCat ion chromatography (IC) system (Metrohm Inc., Switzerland), with method detection limits (MDLs) of 0.20, 0.04 and 0.01 mg

L^{-1}, respectively. Total nitrogen (TN) of samples was measured using Hach Total Nitrogen Reagent set TNT 828 (Hach Inc., US).

5.2.2 Integrated microalgal-bacterial model

The mathematical model was mainly based on the parameters and rates defined by ASM3, which comprises processes of autotrophic bacteria (nitrifiers) and heterotrophic bacteria (denitrifiers). Nitrification and denitrification are represented as single-step processes in ASM3; therefore, modifications were made according to methodology proposed by Iacopozzi et al. (2007) and Kaelin et al. (2009). Nitrification was separated into two processes with NH_4^+ and NO_2^- as substrates for autotrophic bacteria, AOB and NOB respectively. Denitrification was divided into two steps with NO_3^- and NO_2^- as substrates for heterotrophic bacteria (Figure 5.2).

Since algal processes and rates are not accounted for in ASM3, two processes were incorporated, related to algal growth and endogenous algal respiration. Similar to the methodology described by Martinez Sancho et al. (1991), the algae growth was represented by an exponential model, which is one of the most common kinetic models for representing the variability of algae specific growth rate with light intensity:

$$\mu = \mu_{max,P}\left[1 - \exp\left(-\frac{I}{I_S}\right)\right] \tag{5.3}$$

where μ is the algae specific growth rate (d^{-1}), $\mu_{max,P}$ is the maximum specific growth rate for algae (d^{-1}), I is the actual light intensity (μmol photon m^{-2} s^{-1}) and I_S is the saturation light intensity (μmol photon m^{-2} s^{-1}).

The modified Beer-Lambert law was used to incorporate the light intensity variation into the model. Considering equation *(5.2*, it is possible to calculate the point-by-point variation in light intensity inside the PSBR. However, it is very complex to establish this variation in a cylindrical reactor, so an analogy with a parallelepiped was applied to calculate the mean light intensities (I_m) by integrating equation *(5.2* from $x=0$ to $x=L$ (length of the light pathway inside the reactor) and dividing by L:

$$I_m = \frac{I_0}{kX_T L}[1 - \exp(-kX_T L)] \quad (5.4)$$

The average specific growth rate can be described by substituting I_m in equation *(5.3* and assuming that the algal cells adapt to the average value of light intensity and grow as if continuously exposed to that light intensity (Martinez Sancho et al. 1991). Integrating the effect of NH_4^+ substrate concentration (expressed as a Monod equation) and the average light intensity, the algal biomass growth rate is represented by r (g COD m^{-3} d^{-1}):

$$r = \mu_{max,P} \frac{S_{NH_4}}{K_{NH_4,P} + S_{NH_4}} \left\{1 - \exp\left(\frac{-I_o[1 - \exp(-k\,X_T\,L)]}{k\,X_T\,L\,I_s}\right)\right\} X_P \quad (5.5)$$

where S_{NH_4} (g NH_4^+-N m^{-3}) is the NH_4^+-N concentration, $K_{NH_4,P}$ is the NH_4^+ half saturation constant (g NH_4^+-N m^{-3}) and X_P is the phototrophic biomass concentration (g TSS m^{-3}).

The phototrophic endogenous respiration rate, R (g COD m^{-3} d^{-1}), was defined using the same type of mathematical expression as is used for endogenous respiration rates for bacteria as:

$$R = b_P X_P \tag{5.6}$$

where b_P is the endogenous respiration constant for phototrophs (d^{-1}), and X_P is the total solids concentration for phototrophs (g COD m^{-3}).

The mathematical equations were set into the software Aquasim 2.0® (Reichert, 1994) to perform simulations, calibration and sensitivity analysis. The model calibration was done using the data collected hourly during one cycle (24 hours) of R1 on day 49, Phase 2. The initial conditions used as input for the calibration are shown in Table C.1 of Appendix C. The Aquasim tool '*Sensitivity analysis*' was used in order to identify the most sensitive parameters. Afterwards, the calibration was done using the tool '*Parameter estimation*' to estimate new values for the most sensitive parameters, based on the profiles of NH_4^+-N, NO_2^--N, NO_3^--N and DO. The methodology for the sensitivity analysis and calibration is described by Reichert (1998).

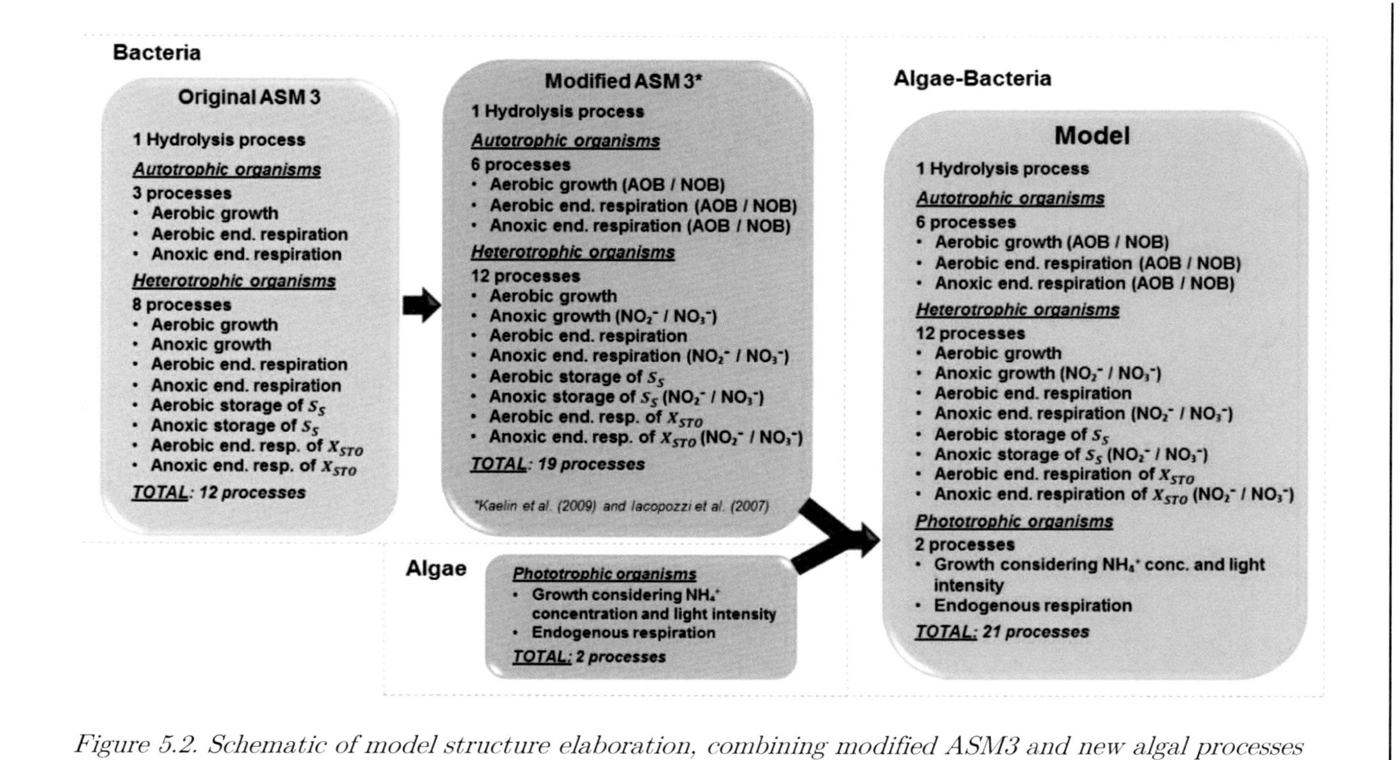

Figure 5.2. Schematic of model structure elaboration, combining modified ASM3 and new algal processes to propose the algal-bacterial model.

5.2.3 Statistical analysis

A statistical analysis applying the *t-test* (two tailed paired) was performed to compare the hourly NH_4^+ removal and NO_2^- formation rates between R1 and R2 during the light period of one cycle. Data from three cycles (Days 14, 42 and 49) were recorded and the average values were used for the statistical analysis. The NH_4^+-N, NO_2^--N and NO_3^--N concentrations in the effluent of R1 and R2 during Phase 1 and Phase 2 were analysed by single factor Analysis of Variance (ANOVA) (α=0.05) using Minitab 16 (PA, USA). The root-mean-square error (RMSE) was used to calculate the error between the values for R1 measured during the experimental period and the values predicted by the model.

5.3 RESULTS AND DISCUSSION

An algal-bacterial consortium was successfully cultured in two PSBRs for 50 days. The biomass developed good settleability with a sludge volume index (SVI) of 53 mL/g for R1 and 58 mL/g for R2. In addition, steady nitritation-denitritation was observed with total nitrogen removal over 90% achieved (see below). Measurements of nitrogen species, biomass concentration and light attenuation were combined with operational parameters to obtain data to calibrate the model.

5.3.1 Experimental

Photo-sequencing batch reactors

Average effluent NH_4^+-N and NO_2^--N concentrations were significantly higher (single factor ANOVA, $p < 0.05$) in Phase 1 than in Phase 2 for both PSBRs (Table

5.2). Total inorganic nitrogen (TIN) removal efficiencies during Phase 1 were approximately 38% and 40% for R1 and R2, respectively. NO_2^- removal by denitrification was most probably hindered by the lack of a readily biodegradable carbon source remaining until the dark period. Likewise, Kinyua et al. (2014) reported that, compared to the total COD of anaerobically digested swine manure, the readily biodegradable COD fraction was very low (4-5%). Wang et al., (2015) showed that little denitritation occurred without addition of an external carbon source when treating anaerobically digested swine manure in a PSBR. Furthermore, previous studies of systems treating wastewater with high levels of total NH_4^+-N and free ammonia have reported inhibition of NOB activity (Kouba et al., 2014; Vadivelu et al., 2007), favouring partial nitrification (i.e. nitritation). Consequently, NO_2^- accumulation was observed in the PSBRs during Phase 1 (Figure 5.3). For this reason, sodium acetate was added to the PSBRs and a 4-day full dark period was implemented to provide conditions required for denitritation. During Phase 2, sodium acetate was added just before the dark cycle to ensure enough readily degradable carbon source for NO_2^- reduction, enhancing TIN removal efficiencies for R1 and R2 to 95% and 94%, respectively.

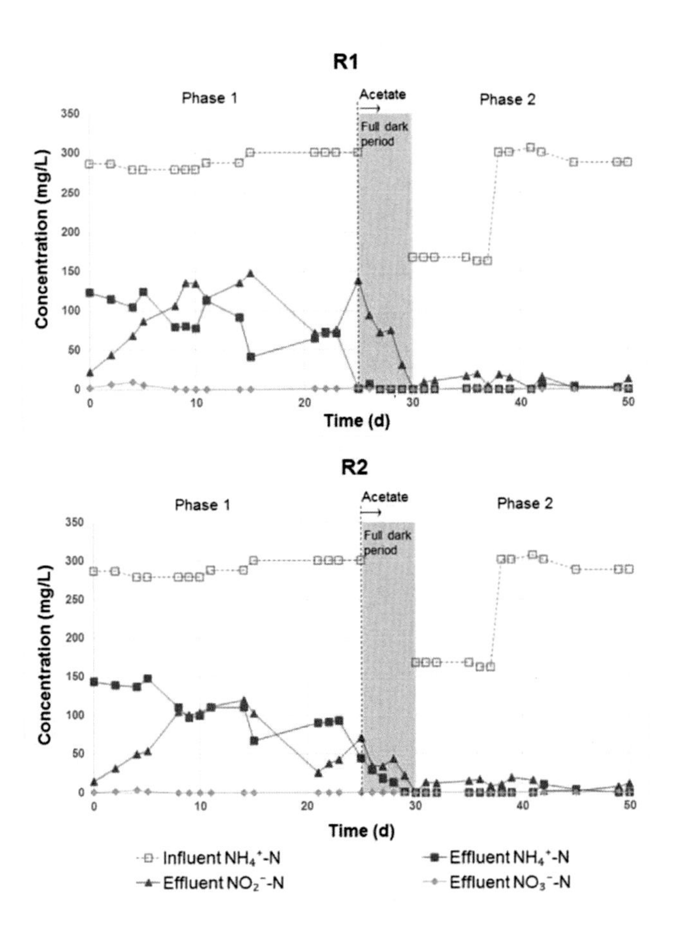

Figure 5.3. Influent and effluent ammonium nitrogen (NH_4^+-N), nitrite nitrogen (NO_2^--N) and nitrate nitrogen (NO_3^--N) concentrations over time in R1 (SRT 7d) and R2 (SRT 11d).

Table 5.2. Average NH_4^+-N, NO_2^--N and NO_3^--N concentrations in the influent and effluent of R1 (SRT 7d) and R2 (SRT 11d). Effluent NH_4^+-N and NO_2^--N concentrations were significantly different between phases, for both reactors. Differences between reactors were not significant (single factor ANOVA 95% confidence interval).

	Influent		***Effluent***					
	R1 and R2		***R1***			***R2***		
	Phase 1	*Phase 2*	*Phase 1*	*Phase 2*	*p value**	*Phase 1*	*Phase 2*	*p value**
NH_4^+-N (mg L^{-1})	290±3	236±19	83±9	1±1	5.14e-12	106±8	5 ± 2	2.73e-14
NO_2^--N (mg L^{-1})	5±0	3±0	97±11	24±7	8.52e-08	70±10	16±3	1.26e-05
NO_3^--N (mg L-1)	< MDL	1 ± 0	2±1	< MDL	-	1±0	< MDL	-

* p value of ANOVA between Phase 1 and Phase 2.

Light attenuation measurements

This study allowed a better understanding of the light attenuation inside the PSBRs and further analysis of how the light attenuation, TSS concentration, oxygen production and nitrogen removal are interlinked. Light intensity varied with distance from the light source inside the reactor and was affected by TSS concentrations (C1 to C7) (Figure 5.4). By fitting Eq. *(5.2)* to these results, the light coefficient k was determined as 0.0748 ± 0.0048 m^2 g^{-1} TSS, later used as an input for the model calibration.

A further analysis based on the light intensities along the light path inside the PSBR at varying TSS concentrations was performed to approximately calculate and compare the portion of irradiated and completely dark volumes in each of the PSBRs. TSS concentrations C5 (1480 mg TSS/L) and C6 (2167 mg TSS/L) were the ones closest to the average in R1 (1357±58 mg TSS/L) and R2 (1744±88 mg TSS/L), respectively. The completely dark volumes were assumed to be the radial portion from the point in which there was no light detected by the quantum meter. For example, for C5 the light intensity was zero from 6cm to 8cm while for C6 the light intensity was zero from 4cm to 8cm distance (Figure 5.).

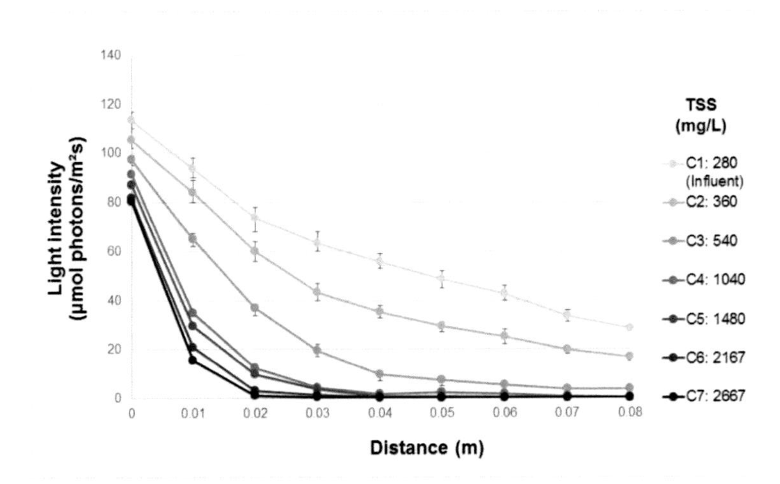

Figure 5.4. Light intensities measured at varying distance from light source inside the PSBR, and varying TSS concentrations (C1-C7).

These values indicate that a higher algal-bacterial biomass concentration hindered the photosynthetic activity in R2 due to the shading effect of the TSS. Therefore not all biomass was continuously irradiated. The average total biomass during the experiment in R2 was 1.44 times higher than in R1. However, applying the percentage of irradiated volume (Figure 5.5) for both reactors and considering only the irradiated biomass, the ratio is almost equalized, lowering the value from 1.44 to 1.09. This indicates that the amount of irradiated algal biomass in both reactors was very similar. Although it is a rough estimation, one can assume that since the oxygen production in the algae chloroplasts is directly related to the light availability, the gross oxygen production for both reactors was similar. The net oxygen production by algae is the gross production minus the oxygen used for algal endogenous respiration. The latter increases with the biomass concentration, and therefore the net oxygen production is probably lower in R2 than in R1. As a result there is more oxygen available for AOB in R1 than in R2 and indeed the NH_4^+ removal and NO_2^- formation rates were significantly higher for R1 than for R2 ($p<0.05$) (Figure 5.6). In addition, the average biomass productivity during the experiment was 187±8 mg L^{-1} in R1 and 156±9 mg L^{-1} in R2. The DO profiles, which are discussed below, also confirmed that more oxygen was available in R1, since the increase in DO towards the end of the light period was higher and started earlier. These results and comparisons indicate that higher SRT resulted in higher TSS concentrations in R2, decreasing the light intensity and oxygen availability for AOB inside the PSBR.

As shown in Figure 5.5, R1 had a higher estimated irradiated volume, due to a lower TSS concentration. It is important to note that these estimations are based on radial decrease of light intensity, while the actual light distribution inside the PSBRs was probably similar to an elliptical shape. This is an artefact of the experimental set-up, in which the light was applied from the sides of the PSBRs. In a full-scale algal pond, light would be coming from the surface of the pond.

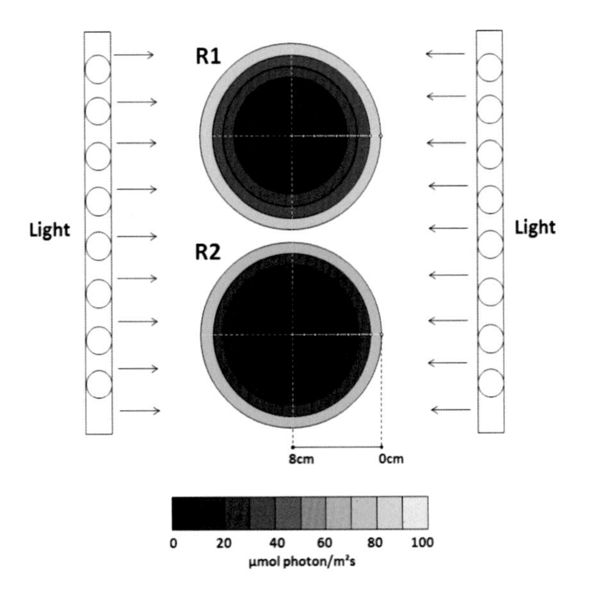

Figure 5.5. Estimation of irradiated zones at varying light intensities (R1: 98%, R2: 75% of reactor volume), and completely dark zones (R1: 2%, R2: 25% of reactor volume) inside both PSBRs.

In photobioreactors with only algal biomass, productivity is maximized when the light intensity is above the compensation light intensity at all locations inside the photobioreactor. Under such conditions all the algal cells are photosynthesizing and there is no dark zone, which increases the biomass productivity (de Mooij et al., 2016). Based on the observed light attenuation in R1, this would require a SRT that is lower than 7 days, to allow further light penetration inside the reactor. Rada-Ariza et al. (2015) observed NH_4^+ removal from 77-96 mg NH_4^+-N L^{-1} in the influent to less than 4 mg NH_4^+-N L^{-1} in the effluent, when the SRT was 3 days or larger. When the SRT was shortened to 1 day, the effluent concentration increased to 18 mg NH_4^+-N L^{-1}. This shows that if the SRT in algal-bacterial systems is too low, slow-growing AOB are washed out of the reactor. Therefore, an optimum SRT should be slightly above the minimum SRT for nitrifiers in order not to decrease the light availability more than necessary. However, as AOB are sensitive to light, the dark zone may also have a secondary benefit as it could protect these microorganisms from photoinhibition (Yoshioka and Saijo, 1984). Furthermore, the presence of a dark zone likely prompted simultaneous nitritation-denitritation during the light period, which was also reported by Wang et al. (2015). This indicates the presence of aerobic and anoxic zones inside the PSBRs in addition to the most probable existence of DO gradients within the algal-bacterial flocs.

In summary, the experiments showed that SRT and light intensity are important factors affecting nutrient removal efficiency in PSBRs, and that SRT should be chosen to optimally balance growth requirements of algae and AOB, since they are combined in one single system.

5.3.2 Integrated microalgal-bacterial model

The list of variables and parameters used in the model, the list of processes and rates and the stoichiometric matrix are provided in Appendix C, table C.2, C.3 and C.4. Profiles of measured values and model predictions of nitrogen species and DO for the light period in both reactors showed a good fit to the experimental data (Figure 5.6). The results for the sensitivity analysis indicated the maximum specific growth rate of phototrophs ($\mu_{max,P}$), saturation constant of NH_4^+ for phototrophs ($K_{NH_4,P}$) and saturation light (I_s) as the most sensitive coefficients for the predictions of nitrogen species and DO. Hence, the calibration resulted in adjusted values for these coefficients (see Table C.2). The following RMSE values were calculated: 8.0 (NH_4^+-N), 6.8 (NO_2^--N), 0.5 (NO_3^--N) and 1.4 (DO). However, the predicted NH_4^+ release during the dark period was significantly higher than observed. An assumption of the model is that only algal 'endogenous respiration' takes place in the absence of light, as well as the bacterial processes. The effect of those processes on NH_4^+ release should be considered. Decostere et al. (2016) proposed a microalgal growth model, which includes respiration and an additional decay process; however, both these processes do not affect the NH_4^+ concentration. In contrast, the heterotrophic respiration taken from ASM3 includes NH_4^+ release (see Table C.5). Apparently the mixed algal-bacterial biomass releases much less NH_4^+ than is observed for endogenously respiring bacteria. This may be explained by the fact that decay and cell disruption are lumped together in ASM3 as 'endogenous respiration'. And decay and disintegration of bacterial cells may occur at higher rates than algal cell disintegration, due to the strong algal cell wall. Therefore, further studies related to the decay and disintegration of algae biomass could elucidate the absence of NH_4^+ release during the dark period (Edmundson and Huesemann, 2015)

The predicted formation and removal of NO_2^- followed the same trend as the experimental data, although the observed decrease in NO_2^- was faster than predicted by the model. This could have been because of an underestimation of the growth rate of denitrifiers, considering that the influent and internally generated COD were ignored. NO_3^- concentrations remained low (<3 mg L^{-1}) throughout the experiments due to the shortcut process of nitritation-denitritation in both experimental data and model performance.

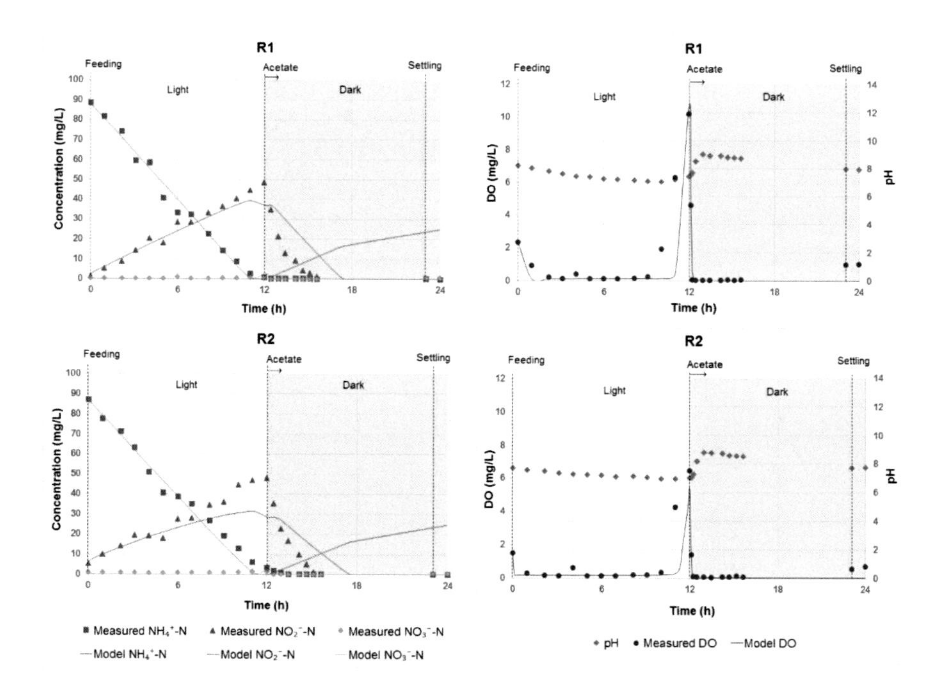

Figure 5.6. Profiles of model predictions and experimental data of nitrogen species and DO for both reactors, during one cycle (Phase 2, Day 49).

In order to compare the performance of NH_4^+ removal in algal-bacterial and algal-only systems, the model was used to simulate PSBR performance with an assumption that R1 contained only algal biomass. This simulation was done using the uncalibrated model (i.e. with parameters from the literature; See Table C.2) and inactivating the bacterial processes in the software (Figure 5.7).

As expected, the simulation did not fit to the observed values, since the AOB and NOB activities were not included. However, the results indicate that NH_4^+ uptake solely by algae in a PSBR occurs at a slower rate than for a mixed consortium of microalgae and nitrifying bacteria (Rada-Ariza et al., 2017). Therefore, the NH_4^+

removal simulated is much lower than the measured values in the PSBRs with the algal-bacterial consortium. The proportion of algae and bacteria in the biomass in R1 was approximately calculated based on the stoichiometry and dry weight obtained from the experiment. The algal-bacterial biomass composition was estimated to be 67% algae, 16% heterotrophs and 17% nitrifiers. The percentage for nitrifiers was similar as observed in a study carried out by van der Steen et al. (2015). The stoichiometric oxidation of NH_4^+ by microbial conversion (nitrification) is much higher compared to the uptake from algal growth. This explains why, even if the algal biomass concentration was much higher than the bacterial biomass concentration, the AOB activity plays an important role in the decrease of NH_4^+ concentration in simulations for combined systems. Hence, when assuming only algal biomass, the NH_4^+ removal is considerably lower than in algal-bacterial systems. In this experiment, the NH_4^+-N removal during one cycle (Phase 2, day 49) was 177 mg NH_4^+-N in R1, from which 96 mg NH_4^+-N (54%) was removed by nitritation-denitritation, and 174 mg NH_4^+-N in R2, from which 87 mg NH_4^+-N (50%) was removed by nitritation-denitritation. Karya et al. (2013) and Wang et al. (2015) reported higher values, with approximately 85% of NH_4^+-N removal in algal-bacterial systems was due to nitrification, and only 15% by algae uptake. It is important to note that the algal performance in this model was only based on NH_4^+ concentration and light availability, but other factors may also be important (i.e. phosphorus concentration, alkalinity and pH).

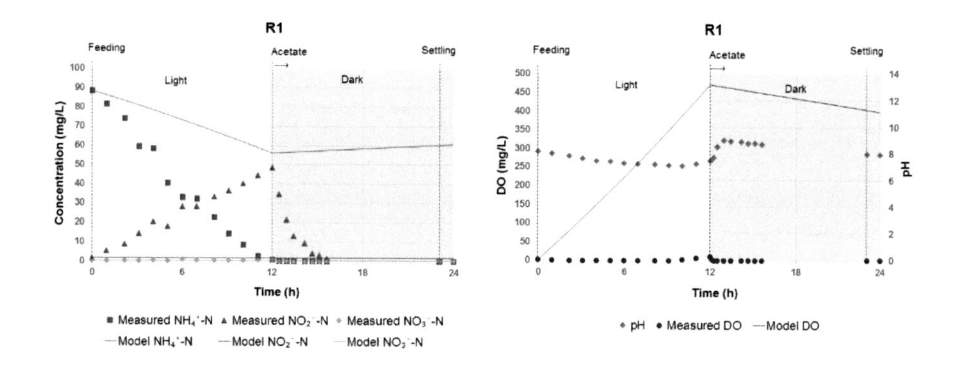

Figure 5.7. Simulations *of the base model (uncalibrated) considering an algal system in R1, i.e. with no bacterial processes incorporated.*

The model presented in this paper, therefore, can help to evaluate nitrogen removal dynamics, as well as to predict the most relevant operating conditions that accelerate or restrict processes in algal-bacterial systems.

5.4 Conclusions

The proposed holistic process has the potential to recover bioenergy from domestic, industrial and agricultural waste while producing treated effluents that can be reused or safely discharged to receiving waters without causing eutrophication. TIN removal (95%) from high NH_4^+ strength wastewater (264 mg NH_4^+-N L^{-1}) using an algal-bacterial consortium in PSBRs was successfully achieved by nitritation-denitritation processes, provided that a biodegradable carbon source was supplied. The operational control of SRT had an important effect on the NH_4^+ removal in the algal-bacterial systems. An SRT of 11 days led to higher TSS concentrations than at SRT of 7 days, hindering the light availability for microalgae due to the self-shading by algal and microbial cells. Consequently, less net oxygen production was observed, decreasing the nitritation rates.

The model developed provided satisfactory results, although further improvements are needed to describe the effect of endogenous respiration on NH_4^+ concentrations during the dark periods of the PSBR cycle. This tool can be useful to design and optimize the operations of PSBRs for different applications (e.g. maximizing algal productivity, minimizing effluent total nitrogen concentration) and different geographic locations and seasons.

6

MODELLING OF NITROGEN REMOVAL USING A MICROALGAL-BACTERIAL CONSORTIUM UNDER DIFFERENT SRTS

This chapter is based on: Rada-Ariza, María, A., Lopez-Vazquez, C.M., Van der Steen, N.P., Lens, P.N.L.. Modelling of the ammonium removal mechanisms in a microalgal-bacterial sequencing-batch photobioreactor. Biotechnology and Bioengineering (Submitted)

Abstract

Mathematical modelling of the microalgal-bacterial consortia was presented in Chapter 5, and shown to be an efficient tool to evaluate the removal mechanisms within the consortia. In this Chapter, further improvement of the model was achieved using the experimental data from Chapter 4. The complete set of experimental data, which included hourly logs of ammonium, nitrite, nitrate, COD and oxygen concentrations at different SRTs, allowed to calibrate and validate the model for different processes. The SRT of 26 days was used for the calibration, while the SRTs of 52 and 17 days were selected for validation. The most sensitive parameters were the maximum growth rate of algae ($\mu_{m,P}$) and light extinction coefficient (k). After calibration and validation, the $\mu_{m,P}$ was found to be 2.00 (± 0.05) d^{-1}, the maximum growth rate of heterotrophic biomass ($\mu_{m,H}$) was 5.5 (± 0.01) d^{-1}, ammonium oxidizing bacteria maximum growth rate ($\mu_{m,AOB}$) was 1.1 (± 0.02) d^{-1} and the maximum growth rate of nitrite oxidizing bacteria ($\mu_{m,NOB}$) was 1.3 (± 0.01) d^{-1}. Furthermore, the minimum SRT for nitrification was calculated using the validated model, and it was found to be between 5 and 10 days. Overall, the model identified the critical point at which the reactor starts to fail, and the limiting conditions when reducing the SRT.

6.1 INTRODUCTION

Microalgal-bacterial consortia can to successfully treat a wide range of wastewater effluents containing different concentrations of nutrients and organic matter, and using different operational conditions (Godos et al., 2010; González-Fernández et al., 2011a; Hernández et al., 2013; Su et al., 2012a; van der Steen et al., 2015; Wang et al., 2015; Liu et al., 2017; Maza-Márquez et al., 2017). The identified removal mechanisms are: nitrification/denitrification, algal uptake, nitrogen requirements for bacterial growth and depending on the pH, ammonium volatilization (Godos et al., 2009; González-Fernández et al., 2011a; Chapter 3; Chapter 4). Microalgal-bacterial systems can generate high quality effluents and due to the photosynthetic oxygenation by algae, the operational costs are expectedly considerably lower compared to conventional wastewater treatment systems (Alcántara et al., 2015).

So far, the removal mechanisms reported for the microalgal-bacterial consortia depend on different conditions, some of the most important are the operational conditions: SRT and HRT, as reported in Chapter 4, and the wastewater characteristics, followed by the reactor design and the environmental conditions (Muñoz and Guieysse, 2006). Considering these conditions, and taking into account the results from Chapter 5, an efficient tool to further study the symbiosis between microalgae and bacteria is the use of mathematical models, which can be useful to get a better understanding of the process, assess and define key operational conditions and eventually can be used to scale up and design a reactor. To date, certain models for microalgal-bacterial biomass have been developed based on calibrated and validated algal models and/or bacterial models (Solimeno et al., 2017; Arashiro et al., 2016; Zambrano et al., 2016; Wolf et al., 2007).

One of the first models for algae and bacteria was proposed by Wolf et al. (2007), the model described the processes occurring in a biofilm composed by chemoautotrophic, photoautotrophic and heterotrophic microorganisms. Furthermore, Decostere et al. (2016) presented a model (developed based on the activated sludge models, ASMs (Henze, 2000)) to describe the activity of the algal biomass. Part of the measured variables and yields were obtained from respirometric experiments, which contributed to the calibration and validation of the model. Solimeno et al. (2017) published the BIO-ALGAE model for microalgal-bacterial biomass growth in high rate algae ponds (HRAP). This model has been calibrated and validated, and has also been proven to be able to provide reasonable predictions of the biomass production.

In Chapter 5, the microalgal-bacterial model described was based on the activated sludge model 3 (ASM3) (Henze, 2000) and the modified ASM3 (Iacopozzi et al., 2007). The microorganisms modelled were ammonium oxidising bacteria (AOB), nitrite oxidising bacteria (NOB), ordinary heterotrophic bacteria (OHO) and photoautotrophic organisms (algae). The model described the nitrification process as a two-step process, and the denitrification process also as a 2-step process. The calibration of the model was carried out based on laboratory experiments using the results of a sequential-batch photobioreactor (described in Chapter 5) that showed that nitratation-denitritation was the main nitrogen removal mechanisms. Furthermore, the authors measured the light extinction coefficient of the biomass in order to take into account the light attenuation effect on the algal growth, following the equation proposed by Halfhide et al. (2015). This helped to calibrate the model using the N-compounds and oxygen concentrations reporting low errors, and therefore, modelling the nitrification and denitrification short-cut processes (nitritation/denitritation). However, the full two step nitrification and denitrification process was not calibrated and overall the model was not validated due to the lack of information.

Therefore, the objective of this chapter is to improve the model proposed in Chapter 5 using the experimental data reported in Chapter 4. The experimental data presents a longer and more complete set of measurements, which includes hourly logs of ammonium, nitrite, nitrate, COD and oxygen concentrations at different SRTs (52, 26 and 17 days). This more complete data set helps to improve the calibration and validation of the model. The main removal mechanism reported in Chapter 4 was via the nitrification/denitrification pathway, which allows to evaluate the two-step approach of these processes proposed in ASM3 and adopted in Chapter 5. In addition, in comparison with the previous version, the available hourly data of COD concentrations allows to evaluate the denitrification and aerobic oxidation of COD, and the estimation of parameters that are sensitive to these processes. The model is calibrated and validated in batch mode and sequential batch mode, allowing to estimate the optimum SRT to maximize the removal rates of the system, and the minimum SRT below which the system starts to fail.

6.2 Materials and Methods

6.2.1 Microalgal-bacterial model

Conceptual model

The model represents the interaction between microalgae and activated sludge microorganisms such as AOB, NOB and OHO. The first version of the model was published by Arashiro et al. (2016), and it is explained in detail in Chapter 5. The model uses the processes and variables defined in the ASM3, and the modified versions of the ASM3 proposed by Iacopozzi et al. (2007) and Kaelin et al. (2009) for the nitrification and denitrification activities, respectively. The modifications included the modelling of nitrification and denitrification activities as two separate processes. The algal activity was modelled by two equations: algal growth and endogenous algal respiration. The algal growth takes into account the light limitation due to the shading effect of the biomass. The model assumes that algal cells can adapt to the corrected average light intensity, when the biomass has reached steady-state conditions. There are in total 21 processes, 16 variables and 47 parameters (Figure 5.2). The nomenclature of each of them, and the Gujer matrix with the stoichiometry and equations are reported in Appendix C.1. The table with the literature values used in this version of the model are presented in Appendix D.1. The software used for the implementation of the model was Aquasim 2.0 (Reichert, 1994), which allows to run the model either in continuous, sequencing-batch mode or batch mode as well as to perform parameter estimations and sensitivity analysis.

The following assumptions were made in order to simplify the model:

1. The growth of algae only takes place using ammonium, as the algal-bacterial biomass did not grow on nitrate based on the results presented in Chapter 4.
2. Phosphorous and alkalinity were not considered limiting factors, therefore the processes related to these variables are not included in this version of the model.
3. The shape of the photobioreactor was considered parallelepipedical instead of cylindrical in order to simplify the calculations of light attenuation (Chapter 5).
4. The aerobic and anoxic processes were active during light and dark phases, and the limiting conditions of the aerobic processes were defined based on the oxygen concentration in the bulk liquid.

6.2.2 Sensitivity analysis

The sensitivity analysis was performed using the sensitivity analysis tool from Aquasim 2.0 (Reichert, 1994). The sensitivity analysis was done using 4 linear sensitivity functions, the one reported in this research was the absolute-relative sensitivity function. This function evaluates the effect of different parameters on specific variables, for this function the units of the parameter do not depend on the units of the variable (Reichert, 1998). The two most sensitive parameters for each of the chosen variables were taken into account for the calibration.

The parameters chosen to perform the sensitivity analysis were: the maximum growth rates of AOB ($\boldsymbol{\mu_{m,AOB}}$), NOB ($\boldsymbol{\mu_{m,NOB}}$), OHO ($\boldsymbol{\mu_{m,H}}$), and phototrophs ($\boldsymbol{\mu_{m,P}}$), the light extinction coefficient($\boldsymbol{k}$), the light saturation constant (I_s), and the COD storage rate constant (k_{STO}). Some of the parameters were chosen based on the results obtained by Solimeno et al. (2017), Decostere et al. (2016) and

Zambrano et al. (2016) and the sensitivity analysis performed in Chapter 5. The effect of these parameters was evaluated on the variables of ammonium, nitrite, nitrate, COD and oxygen concentrations.

6.2.3 Reactor and data collected

The microalgal-bacterial biomass characteristics and the reactor used for this model are described in detail in Chapter 4. The reactor was operated as a sequencing batch reactor, and each cycle had a duration of 12 hours (HRT). In each of the cycles there were two light and two dark phases, it was reported that the light phases were considered to be aerobic phases and the dark phases anoxic phases. The reactor was operated for 300 days, and during this time 4 different SRTs were tested: 48, 52, 26 and 17 days. For the calibration and validation of this model the SRT of 52 (period 1A), 26 (period 1B), and 17 (period 1C) days will be used, since these three periods have the same feeding and operational conditions (as described in Chapter 4).

In each of the periods a detailed sample collection of some cycles was carried out. Samples were taken every half an hour to measure N-nitrogen compounds and COD. In addition, O_2 measurements were recorded every 5 seconds using an O_2 probe, and samples for the analysis of Chlorophyll-*a*, VSS and TSS concentrations were collected for every cycle. The information of periods 1A, 1B and 1C was used to calibrate and validate this model. In Chapter 4, a detailed example of the information collected and the number of cycles per period is described. This detailed information was used to calculate the nitrogen ammonium removal rates, oxygen production and biomass characterization (also presented in Chapter 4).

6.2.4 Calibration and validation of the microalgal-bacterial model

The calibration of the microalgal-bacterial model was done using the measured data of period 2B. The calibration was carried out for two operational modes: (i) batch mode and (ii) sequencing batch mode. Period 2B was chosen to calibrate the model because it has the highest amount of measured cycles. The parameters to calibrate were chosen based on the sensitivity analysis. The half-saturation constants of ammonium ($\boldsymbol{K_{NH4,AOB}}$) and oxygen ($\boldsymbol{K_{O2,AOB}}$) for AOB, and the half-saturation ammonium constant for phototrophs ($\boldsymbol{K_{NH4,P}}$) were also calibrated. The rest of the parameters was not calibrated and remained as the typical values reported in the literature (Table D.1, Appendix D.1).

Calibration and validation of the batch mode

The calibration of the batch operational mode did not take into account the waste of sludge, and it was performed using the parameter estimation tool from Aquasim. The model calibration of the batch mode was performed in two steps: first, the variables of ammonium, nitrite, and nitrate and oxygen concentration were fitted to the detailed measured data of the cycles by calibrating the following parameters: $\mu_{m,AOB}$, $\mu_{m,NOB}$, $\mu_{m,H}$, $\mu_{m,P}$, k, I_s, $K_{NH4,AOB}$, $K_{O2,AOB}$, and $K_{NH4,P}$. The second step was to calibrate the k_{STO} using the measured data of COD. Other input parameters of the model, such as the fraction of AOB, NOB and algae in the biomass were determined previously (as presented in Chapter 4), and were included in the model as initial biomass composition. These values were calculated based on mass balances for the nitrogen removal of each group of microorganisms (bacteria and algae), their stoichiometry, and the VSS measured in the reactor. The values were introduced in the model in units of mg COD L^{-1}. Thus, in the case of bacteria (OHO and nitrifiers), a conversion factor of 1.48 mg COD mg VSS^{-1} was used (Ekama and Wentzel, 2008b), and for algae biomass the factor was 0.953 mg COD mg VSS^{-1} (Zambrano et al., 2016).

In order to validate the model, the values of the parameters calibrated in period 2B were used in the remaining periods (2A and 2C). As such, for each period, the initial characteristics of the biomass and the initial concentrations of ammonium, nitrate, nitrite and COD, were the only input data modified. The results from the calibration and validation of the model were compared with the data measured and the errors between predicted and observed data were calculated to assess the fitting of the model. The results of the calibration and validation are presented in Appendix D.2.

Calibration of the sequencing batch mode operation

The operation of the reactor in the laboratory in a sequencing batch mode was composed of cycles that included the feeding period, reaction time (lights on and off), settling phase and effluent withdrawal (Figure 4.1). Therefore, the model in Aquasim was also set up in a sequencing batch mode following the operation of the lab system (including the influent addition, sludge waste and effluent withdrawal, and the lights turned on and off). The SRT was defined setting a defined volume of waste of sludge per day.

Figure 6.1 shows the conceptual tanks defined in Aquasim to represent the operation of the reactors. There were three tanks defined in the model: (i) mixed reactor, (ii) waste sludge tank (WAS tank), and (iii) the effluent tank. The sedimentation was not defined within the cycle of the reactor in Aquasim, but modeled using defined biomass retention ratios from the effluent. Therefore, part of the biomass would remain in the reactor (f_Xret_i) and the rest would leave the reactor through the effluent ($1 - f\_Xret_i$). Prior to the calibration of the removal rates and trends of N-compounds, COD and oxygen concentrations, the calibration of the modelled biomass was done in Aquasim. The calibration of the biomass was perfomed by adjusting the retention ratios from the effluent to the reactor. The recirculation takes place at the end of the reaction time, and within the model the process is immediate. Given that the settling properties of the biomass differed among microbial populations, besides the definition of the SRT, different retention ratios were defined for algae (f_Xret_P) and bacteria (ammonium oxidising bacteria: f_Xret_{AOB}, nitrite oxidising bacteria: f_Xret_{NOB}, heterotrophic bacteria: f_Xret_H and inert solids f_Xret_I).

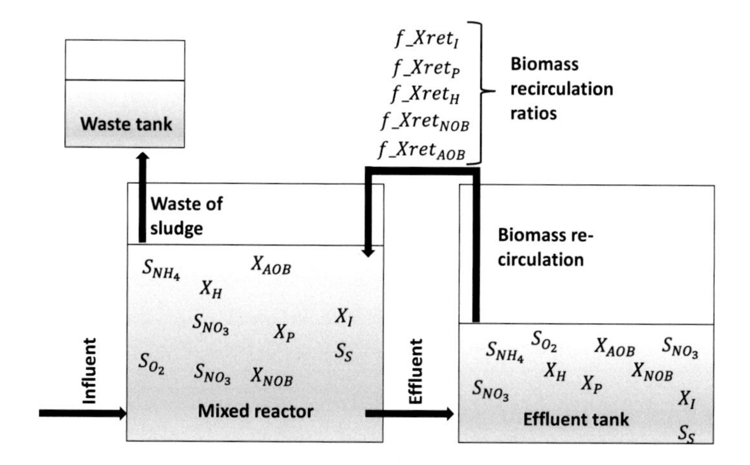

Figure 6.1. Graphical scheme of the conceptual arrangement of tanks of the photo-activated bioreactor in Aquasim.

The modelled biomass and fractionation was compared with the measured VSS reported in Chapter 4, and also the fractionation of the biomass into the different microorganisms: heterotrophic, ammonium and nitrite oxidizing bacteria and microalgae. The calibration of the ratios was carried out with data from period 2B and validated with data from periods 2A and 2C. The modelled biomass concentration was compared with the average measured VSS of each period.

Once the calibration and validation of the biomass was achieved, the calibration of the concentrations of ammonium, nitrite, nitrate, COD and oxygen was done manually. First, selecting the parameters to calibrate based on a sensitivity analysis. In addition, taking into account that the kinetics between the batch mode and the sequencing batch mode operation differed, the re-calibration of the following parameters was necessary: $\mu_{m,P}$, $\mu_{m,AOB}$ $\mu_{m,NOB}$, $\mu_{m,H}$, and k. Basically, the re-calibration consisted of the growth rates of the different microorganisms and the light extinction coefficient. The other parameters remained the same as calibrated in the batch mode.

In order to reach steady-state conditions in the model for the sequencing batch mode of operation, for each of the periods the model was run for an equivalent duration of 6 times the SRT applied. Thus, in the case of the SRT of 26 days, the model ran for 156 days. Thereafter, the comparison between the calibrated model and the measured data of the cycles was done using the last days of the modelling, ensuring that it reached steady-state conditions.

The validation of the model was performed using the data measured in periods 2A and 2C. The light attenuation coefficient (k) was the only parameter that was modified for each of the periods during the validation.

6.2.5 Calculation of the error

The index of agreement (IOA) *(6.1)* (Wilmot et al., 2012) was calculated in order to evaluate the error of the model in comparison with the measured data. The index of agreement estimates the variance of the model and compares the results with the variance of the measured or observed data. Therefore, when the result is equal to 1 it means that the variance of the model is lower than the variance of the data. When the index IOA is equal to zero, the variance of the model is equal to the variance of the observed data. Lastly, when the IOA is lower than zero means that the variance of the model is higher than the variance of the observed data, this is a negative indication of the performance of the model. Any value between zero and 1 is acceptable. Furthermore, the closer to 1 the better the description of the model of to the real data.

$$IOA = 1 - \frac{\sigma\varepsilon}{\sigma obs} = 1 - \frac{\sum(y_{obs} - y_{mod})^2}{\sum(y_{obs} - \overline{y_{obs}})^2} \tag{6.1}$$

Where:

y_{obs}: Measured data points.

y_{mod}: Modelled data points.

$\overline{y_{obs}}$: Average of the measured series.

6.2.6 Evaluation of shorter SRTs

The shortest SRT at which the photobioreactor was operated during the experimental phase was 17 days. Therefore, in order to further investigate the outcome of the microalgal-bacterial system at SRT shorter than 17 d, once the model was calibrated and validated, it was run in a sequencing batch mode operation using shorter SRTs. The objective was to identify the minimum SRT at which nitrification/denitrification stops, which affects the performance of the system for N removal and ultimately the failure of the system.

The SRTs modelled were firstly 15, 10, and 5 and subsequently 3, 2, 1 and 0.8 days. The approach to model all proposed SRTs was as follows:

- In order to ensure that steady-state conditions were achieved, for every SRT tested the model ran for a duration of 120 days. In most cases, the steady-state was achieved at a duration of 6 times the SRT applied; for instance, for the SRT of 15 days, steady-state conditions were achieved after an equivalent simulation time of 90 days.
- Prior to modelling the scenarios, the model was calibrated with the SRT of 26 days and validated with the SRT of 17 and 52 days.
- After calibration and validation, the SRT was changed in the model to the already defined values and the results of the concentrations of ammonium, nitrate, nitrite, oxygen, organic carbon and biomass from the days 98 to 100 were reported and compared.

6.3 Results and discussion

6.3.1 Sensitivity analysis

The sensitivity analysis was performed to identify the effect of different parameters on the following variables: ammonium, nitrite, nitrate, COD and dissolved oxygen concentrations (Figure 6.2). The parameters selected were the maximum growth rates of AOB ($\boldsymbol{\mu_{m,AOB}}$), NOB ($\boldsymbol{\mu_{m,NOB}}$), OHO ($\boldsymbol{\mu_{m,H}}$), and phototrophs ($\boldsymbol{\mu_{m,P}}$), the light extinction coefficient ($\boldsymbol{k}$), the light saturation constant ($\boldsymbol{I_s}$), and the COD storage rate constant ($\boldsymbol{k_{STO}}$). In the previous version of this model (Chapter 5), the sensitivity analysis was performed using fewer parameters ($\boldsymbol{\mu_{m,P}}$, $\boldsymbol{I_s}$, $\boldsymbol{K_{NH4,P}}$), which were identified to be the most sensitive. For this version, since the two-step nitrification step is modelled, the growth rates of AOB and NOB were included. Also, the growth rate of heterotrophs was added, since it defines the rate of denitrification and COD oxidation.

Figure 6.2 shows the results of the sensitivity analysis, the effect of the different parameters on each of the variables of interest was evaluated during the reaction time of the cycle (5 hours). For each of the variables, the concentrations calculated show the dependence of the sensitivity functions with regard to each parameter. The two most sensitive parameters for each variable were chosen for calibration. The sign (positive or negative) of the concentration defines if the parameter has a direct or inverse proportional effect on the variable. For instance, negative concentrations mean that the concentration of the variable decreases as the parameter increases, and the opposite when the variable increases.

Two of the parameters that were identified to be the most sensitive were $\boldsymbol{\mu_{m,P}}$ and $\boldsymbol{I_s}$ (Figure 6.2). These two values had the strongest effect on all the variables modelled. Accordingly, Decostere et al. (2016) and Solimeno et al. (2017) reported the maximum growth rate of algae as one of the most sensitive parameters for the algal and algal-bacterial model, respectively. Zambrano et al. (2016) reported that the ammonium and oxygen concentrations were most sensitive to the $\boldsymbol{\mu}$ of algae and bacteria.

In this research, besides the $\boldsymbol{\mu_{m,P}}$ and the $\boldsymbol{I_s}$, other sensitivity estimations of more parameters were analysed in order to contribute to explain the different relationships between these ones and the variables modelled. The ammonium concentration (Figure 6.2.A) is affected by $\boldsymbol{\mu_{m,P}}$, $\boldsymbol{\mu_{m,AOB}}$, $\boldsymbol{\mu_{m,H}}$, and an increase of these parameters causes a decrease in the ammonium concentration, due to a higher consumption of ammonium for biomass growth and/or oxidation.

In the case of nitrite (Figure 6.2.B) and nitrate (Figure 6.2.C) concentrations, the $\boldsymbol{\mu_{m,P}}$ and $\boldsymbol{\mu_{m,AOB}}$ have the strongest effect. An increase in $\boldsymbol{\mu_{m,P}}$ results in a decrease in the NO_2^--N concentrations and an increase of the NO_3^--N concentrations, since there is more oxygen available to fully complete the two-step nitrification. The opposite occurs when there is an increase in $\boldsymbol{\mu_{m,AOB}}$, the nitrite concentration increases while the nitrate concentration decreases. On the contrary, an increase in the NOB growth rate ($\boldsymbol{\mu_{m,NOB}}$) causes a decrease in the nitrite concentration and an increase in the nitrate concentration. The growth rate of OHO only has an effect on the nitrate concentration, which decreases when there is an increase in $\boldsymbol{\mu_{m,H}}$, which is attributed to a higher denitrification potential.

One parameter that was found to have a strong effect on the concentration of COD (Figure 6.2.D) was the COD storage rate constant ($\boldsymbol{k_{STO}}$), an increase in this parameter decreases the COD concentration in the reactor. The same effect is caused by an increase in $\boldsymbol{\mu_{m,P}}$, as there is more oxygen at a higher microalgal growth rate. As expected, the oxygen concentration (Figure 6.2.E) was highly affected by the $\boldsymbol{\mu_{m,P}}$, and the $\boldsymbol{I_s}$.

Based on the sensitivity analysis, the parameters chosen for calibration were $\boldsymbol{I_s}$, $\boldsymbol{k}$, $\boldsymbol{\mu_{m,AOB}}$, $\boldsymbol{\mu_{m,NOB}}$, $\boldsymbol{\mu_{m,P}}$, $\boldsymbol{\mu_{m,H}}$ $\boldsymbol{k_{STO}}$ and $\boldsymbol{k}$. The last parameter (light extinction coefficient) was chosen for calibration despite that it had a minimal or almost negligible effect on the rest of the variables during the time tested. The light extinction coefficient is a physical characteristic of the biomass that was not measured during data collection. Hence, it needed to be estimated by the model. Furthermore, other parameters selected for calibration that were not included in the sensitivity analysis were $\boldsymbol{K_{NH4,AOB}}$, $\boldsymbol{K_{O2,AOB}}$, $\boldsymbol{K_{NH4,P}}$. The selection of these parameters for calibration was based on the previous results in Chapter 5.

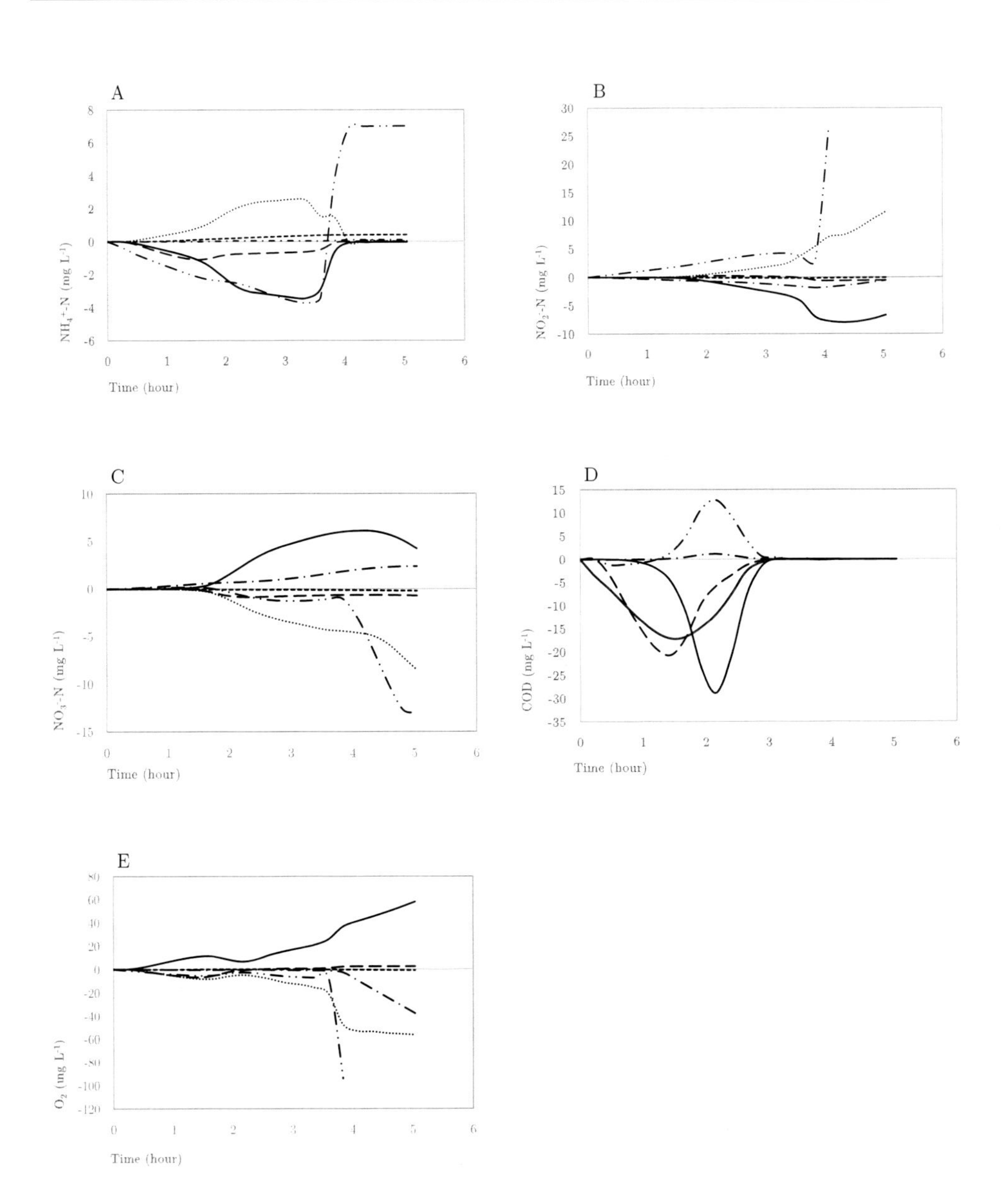

Figure 6.2. Sensitivity analysis of the mathematical prediction of (A) NH_4^+-N, (B) NO_2^--N, (C) NO_3^--N, (D) COD and (E) O_2, with respect to: $\boldsymbol{I_s}$ *(.........),* $\boldsymbol{k}$ *(----·),* $\boldsymbol{\mu_{m,AOB}}$ *(— · ·),* $\boldsymbol{\mu_{m,NOB}}$ *(— · -),* $\boldsymbol{\mu_{m,H}}$ *(— —),* $\boldsymbol{\mu_{m,P}}$ *(——),* and $\boldsymbol{k_{STO}}$ *(═══).*

6.3.2 Calibration and validation of the N-compounds, oxygen and COD in batch operational mode

The calibration of the total concentration and composition of the algal-bacterial biomass was done using the laboratory data of period 1B. The parameters calibrated (Table D.2) in the first part were I_s, k, $K_{NH4,AOB}$, $K_{O2,AOB}$, $K_{NH4,P}$, $\mu_{m,AOB}$, $\mu_{m,NOB}$ and $\mu_{m,P}$. These were calibrated by fitting the model to the measured concentrations of ammonium, nitrate, and O_2. The k_{STO} was calibrated in the second part of the calibration, using the COD concentration data since it was one of the most sensitive parameters with regard to the organic carbon concentration. The fitting between the measured data and the data predicted by the model for period 2B is shown in Figure 6.3. The error between the modelled and measured data was determined by the IOA, once the model was calibrated. The IOA calculated were 0.91, 0.89, 0.73 and 0.69 for the ammonium, nitrate, COD and oxygen concentrations, respectively. Therefore, the variance of the model is lower than the variance of the observed data. The model closely describes the laboratory data for period 2B, and the processes, coefficients and variables proposed and defined in the model could follow the trend of the different variable concentrations of the runs in a batch operational mode.

The calibrated parameters using the data from period 2B were successfully validated for periods 2A and 2C. The microalgal-bacterial model showed a satisfactorily description of the concentrations of ammonium and nitrate, but less accurate for COD and O_2. Nevertheless, the description of the measured data can be considered acceptable. Figures D.1 and D.2 show the modelled and measured data for ammonium, nitrate, COD and oxygen concentrations. The values of the IOA are presented in the graphs (Figure D.1 and D.2).

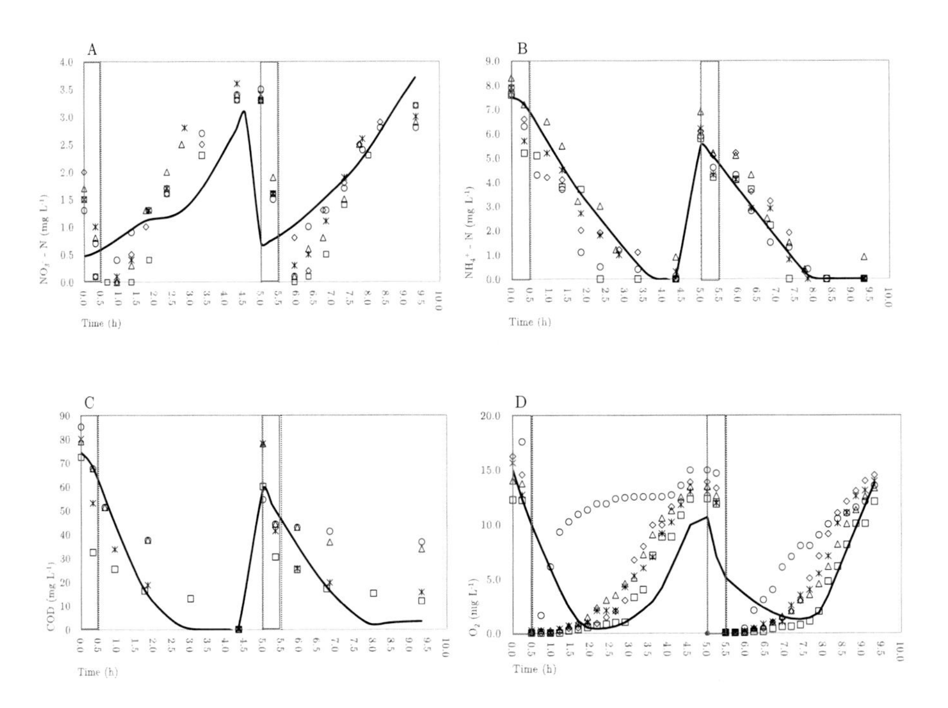

Figure 6.3. Modelled and measured data for NO_3^--N (A), NH_4^+-N (B), COD (C) and O_2 (D) concentrations during period 1B. Solid line: model data during period 1B; measured data during period 1B in cycles C1 (□), C2 (◇), C3 (○), C4 (△) and C5 (*).

6.3.3 Calibration and validation of the biomass characterization and production in sequencing batch mode operation

To simulate the settling properties of the different biomasses, certain retention ratios for the different biomass species were defined during the calibration in period

1B (Figure 6.4). Thus, the retention ratios were found to be 0.7 for the heterotrophic biomass (f_Xret_H), ammonium oxidising (f_Xret_{AOB}) and nitrite oxidising (f_Xret_{NOB}) biomass; and 0.6 for the microalgal biomass (f_Xret_P). The validation was performed with data from periods 1A and 1C (Figure D.3).

In Chapter 4, the composition of the biomass into the different groups of microorganisms was estimated based on the stoichiometry of their biomass composition (Figure 4.5). This information was compared with the results from the model (Table 6.1). In the case of period 1B, the model could describe accurately not only the total VSS concentration of the biomass, but also the fractions of the different microorganisms present ($p>0.05$). For validation purposes, the model predictions of the total VSS concentrations of period 1A were not significantly different than the measured data ($p>0.05$). However, this was not the case for period 1C ($p<0.05$), in which the total predicted biomass concentration is significantly higher than the measured data. Possibly, the differences during period 1C might be due to the recirculation factors, yet the retention ratios were not modified for this period. One of the reasons is that the predictions of the concentrations of the dissolved parameters (NH_4^+, NO_3^-, NO_2^-, COD and O_2) led to low errors. Also, the difference between the measured and predicted concentrations of the biomass was not higher than 40%. Furthermore, for the three periods of study, the concentrations of nitrifiers described by the model were lower than the values estimated theoretically using the stoichiometry and measured data (Table 6.1). Overall, the defined biomass yields, used in the model and selected from literature, could describe the biomass production and fractionation within the microalgal-bacterial reactor (Figure 6.4).

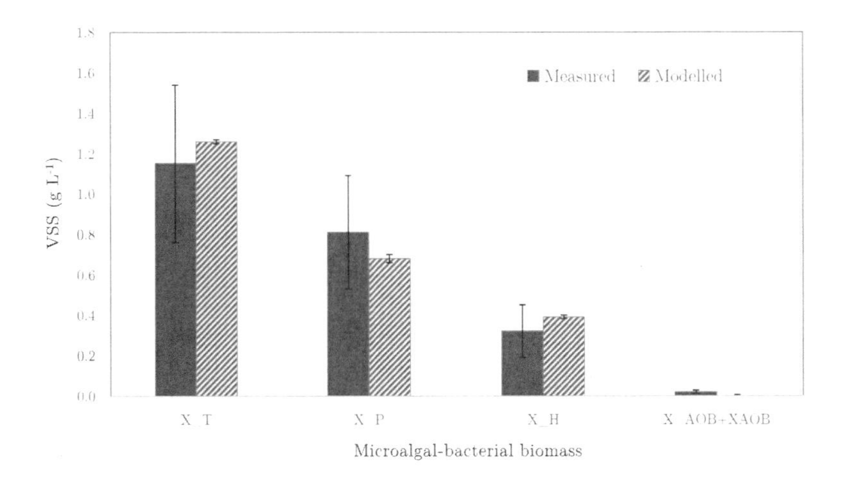

Figure 6.4. Comparison between the modelled and measured biomass in the microalgal-bacterial reactor for period 1B. X_T: Total biomass, X_P: Phototrophic biomass, X_H: Heterotrophic biomass, X_AOB+X_XNOB: Ammonium and nitrite oxidising bacteria.

Furthermore, in the case of period 1C, it can be observed that the heterotrophic biomass concentration is higher than the phototrophic biomass concentration, which was well described by the model. This can be attributed to the shorter SRT and the lower retention ratio of algae (0.6) in comparison with that of the bacterial biomass (0.7). In addition, the growth rate of the heterotrophic bacteria (5.5 ± 0.01 d^{-1}) is higher than the algae (2.00 ± 0.05 d^{-1}) (Table 6.2). For periods 1B and 1A (longer SRTs), algal biomass represented on average 50% of the biomass and the rest was heterotrophic bacteria and inert biomass, while nitrifiers only comprised a very small fraction of the total mass (0.3-0.7%). Similar to these observations, Solimeno et al. (2017) reported that in a modelling study algae represented between 58 and 68.4% of the total biomass, whereas nitrifiers only between 0.15 and 0.18%.

The smaller retention factor of the algal biomass is in accordance with the physical characteristics of algal cells, and their poor settleability, when compared with activated sludge bacteria. Algal cells commonly are smaller than 30 μm, and the settling velocities are not higher than 10^{-6} m s^{-1} (Granados et al., 2012). However, based on the experimental data (Chapter 4), the agglomerates made by algae and bacteria helped to increase the biomass settleability.The higher settleability due to the formation of agglomerates has been also reported by other authors (Quijano et al., 2017), which is enhanced under sequencing batch reactor operational modes through the selection of the fastest settling species (de Godos et al., 2014; Van Den Hende, 2014; Valigore et al., 2012). With regard to the model, special assumptions and considerations regarding the retention ratios should be made when the biomass is composed of activated algal granules. Algal granules have higher settling velocities (21.6 ± 0.9 m h^{-1}) than algal-bacterial agglomerates or flocs (Tiron et al., 2017).

Table 6.1. Biomass concentrations based on mass balances (Exp.) in the microalgal-reactor and modelled biomass (Model) using the microalgal-bacterial model.

Biomass	*Biomass concentrations ($gVSS\ L^{-1}$)*					
	1A	*1A*[2]	*1B*	*1B*[1]	*1C*	*1C*[2]
	Exp.	*Model*	*Exp.*	*Model*	*Exp.*	*Model*
X_P	1.95±0.23	1.42±0.00	0.81±0.28	0.68±0.02	0.34±0.14	0.53±0.00
X_H	0.63±0.13	0.74±0.00	0.32±0.13	0.39±0.00	0.53±0.07	0.84±0.00
X_{AOB}, X_{NOB}	0.05±0.00	0.01±0.00	0.02±0.00	0.01±0.00	0.06±0.04	0.01±0.00
X_T	2.64±0.30	2.80±0.00	1.15±0.39	1.26±0.01	0.92±0.22	1.43±0.01

[1]Calibrated period

[2]Validated periods

6.3.4 Calibration and validation under sequencing batch operational mode of then concentrations of N-compounds, oxygen and organic carbon

The calibration of the model at the SRT of 26 days (period 1B) for the sequencing batch mode operation was carried out for an equivalent duration of 150 days. The stable conditions in the model were reached after day 120. The parameters calibrated for the batch and sequencing batch operational modes are presented in Table 6.2. The results of the modelled concentrations of N-compounds between day 130.5 and day 131 are presented in Figure 6.5, while Figure 6.6 presents the comparison between the modelled (steady state) and average measured parameters for the cycles (ammonium, nitrite, nitrate, oxygen and COD) and the calculated IOA.

Table 6.2. Calibrated parameters for the algal-bacterial model and literature values.

Symbol	*This study*	*Unit*	*Typical values reported in literature*	*Reference*
I_s	35.0 ± 0.4[1]	μmol photon m^{-2} s^{-1}	13	Martinez Sancho et al. (1991)
			758 ± 23	Wágner et al. (2016)
k	0.019 ± 0.003[2]	m^2 $gTSS^{-1}$	0.07	Molina Grima et al. 1994; Solimeno et al. (2017)*
			0.0748	Arashiro et al. (2016)*
			0.29-0.25	Blanken et al. (2016)
$K_{NH4,AOB}$	0.13 ± 0.02[1]	g N m^{-3}	0.5	Solimeno et al. (2017)*; van der Steen et al. (2015)*; Reichert et al. (2001)
			2	Iacopozzi et al. (2007); Henze (2000)
			2.4	Chapter 5*; Wiesmann (1994)
$K_{O2,AOB}$	0.75 ± 0.01[1]	g O_2 m^{-3}	0.5	Solimeno et al. (2017)*; Reichert et al. (2001); Henze (2000)
			0.79	Chapter 5*; Manser et al. (2005)

Symbol	*This study*	*Unit*	*Typical values reported in literature*	*Reference*
$K_{NH4,P}$	0.001 ± 0.000[1]	g N m^{-3}	0.00021	Chapter 5*
			0.017	Wolf et al. (2007)*
			0.1	Solimeno et al. (2017)*; Reichert et al. (2001)
			0.1	Zambrano et al. (2016)
			0.3	Decostere et al. (2016)
			2.13 ± 0.86	Wágner et al. (2016)
$\mu_{m,AOB}$	1.10 ± 0.02[2]	d^{-1}	0.11	van der Steen et al. (2015)*
			0.63	Gujer et al. (1999); Solimeno et al. (2017)*
			0.6313	Iacopozzi et al. (2007)
			0.9	Chapter 5*; Kaelin et al. (2009)
$\mu_{m,NOB}$	1.30 ± 0.01[2]	d^{-1}	0.5	van der Steen et al. (2015)*
			0.65	Chapter 5*; Kaelin et al. (2009)
			1.0476	Iacopozzi et al. (2007)
			1.1	Gujer et al. 1999; Solimeno et al. (2017)*

Symbol	*This study*	*Unit*	*Typical values reported in literature*	*Reference*
$\mu_{m,P}$	2.00 ± 0.05^2	d^{-1}	0.13	Choi et al., (2010)
			0.15-0.39	Decostere et al. (2016)
			0.62	van der Steen et al. (2015)*
			0.85	Chapter 5*
			1.6	Zambrano et al. (2016)*
			1.5	Solimeno et al. (2017)*
			2.37	Martinez Sancho et al. (1991)
			3.6±0.04	Wágner et al. (2016)
			6.48 and 3.36	Blanken et al. (2016)
$\mu_{m,H}$	5.5 ± 0.01^2	d^{-1}	2	Henze (2000)
			1.3	Solimeno et al. (2017)
			1	Chapter 5*
			2-10	Metcalf & Eddy et al. (2002)
k_{STO}	0.88 ± 0.03^1	g COD_{Ss} g COD_{XH}^{-1} d^{-1}	5	Chapter 5*-; Henze (2000)
			7.38	Iacopozzi et al. (2007)

*. Microalgal-bacterial biomass models.

[1]. Batch operational mode.

[2]. Values calibrated under sequetial batch mode operation.

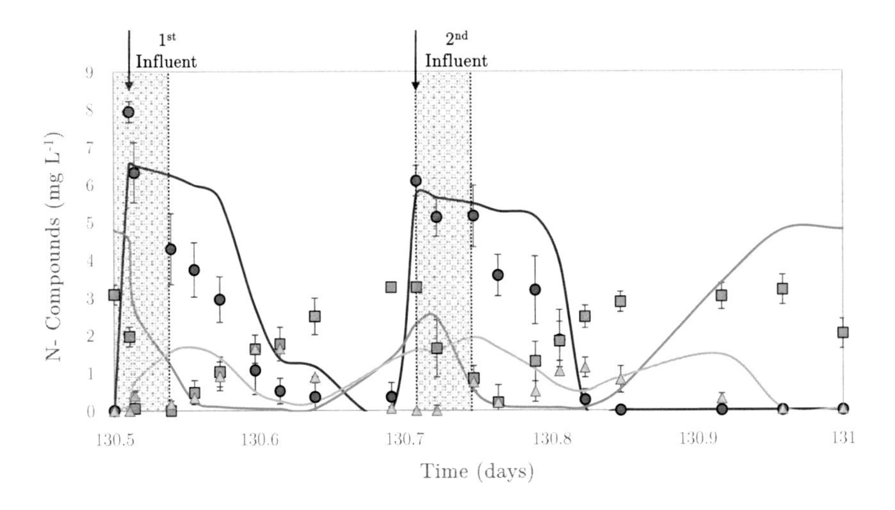

Figure 6.5. Calibrated N-compounds concentrations for period 1B with a SRT of 26 days: NH_4^+-N (▬), NO_3^--N (▬) and NO_2^--N (▬) modelled concentration and NH_4^+-N (●), NO_3^--N (■) and NO_2^--N (▲) measured concentration (average between of the measured cycles). The grey-shaded areas correspond to the dark phases (lights turned off) during the cycles.

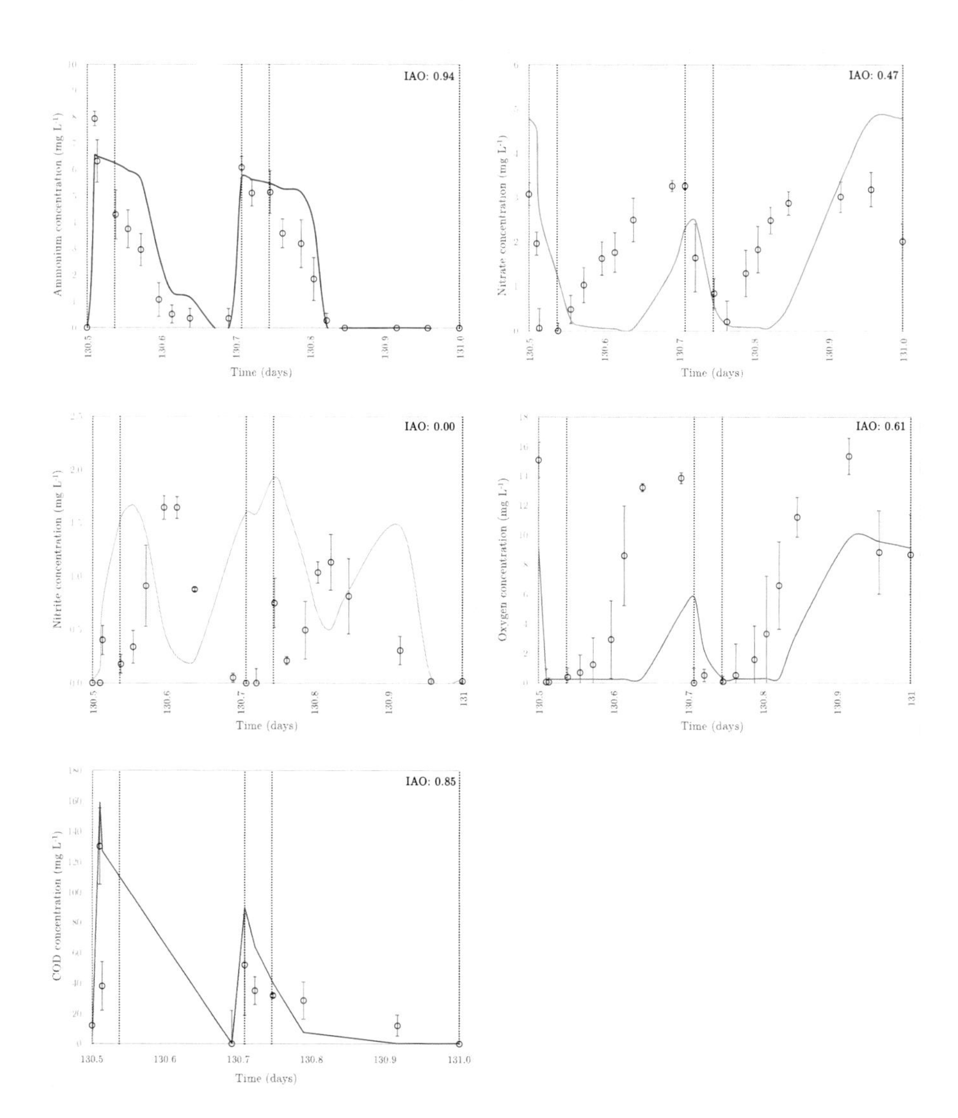

Figure 6.6. Description of the concentrations of ammonium (▬), nitrate (▬), nitrite (▬), oxygen (▬), COD (▬) for period 1B with a SRT of 26 days after calibration and comparison with the measured data (-o-)(average between of the measured cycles).

The validation of the model under sequencing batch mode operation was done using the remaining periods 1A and 1C (Figure 6.7). However, the different SRTs had an impact on the solids concentration, which ultimately affects the light availability for algae and as such the increase or decrease of the dark zones (under complete mixed conditions). Therefore, in order to model the effect of the biomass concentration on the light availability, the light attenuation coefficient ***k*** was adjusted in periods 1A and 1C. For the calibrated period 1B (26 days SRT), ***k*** had a value of 0.020 m^2 $gTSS^{-1}$, while for periods 1A (52 days) and 1C (17 days), ***k*** was 0.015 and 0.021 m^2 $gTSS^{-1}$, respectively. Thus, the average light extinction coefficient of the biomass was estimated around 0.019 ± 0.003 m^2 $gTSS^{-1}$.

Table 6.3 IOA calculated for the modelled parameters in period 1A and 1C.

Parameter	*IOA*	
	Period 1A	*Period 1C*
$NH_4^+ - N$	0.88	0.81
$NO_2^- - N$	0.16	0.00
$NO_3^- - N$	0.20	0.44
O_2	0.66	0.76
COD	0.53	0.86

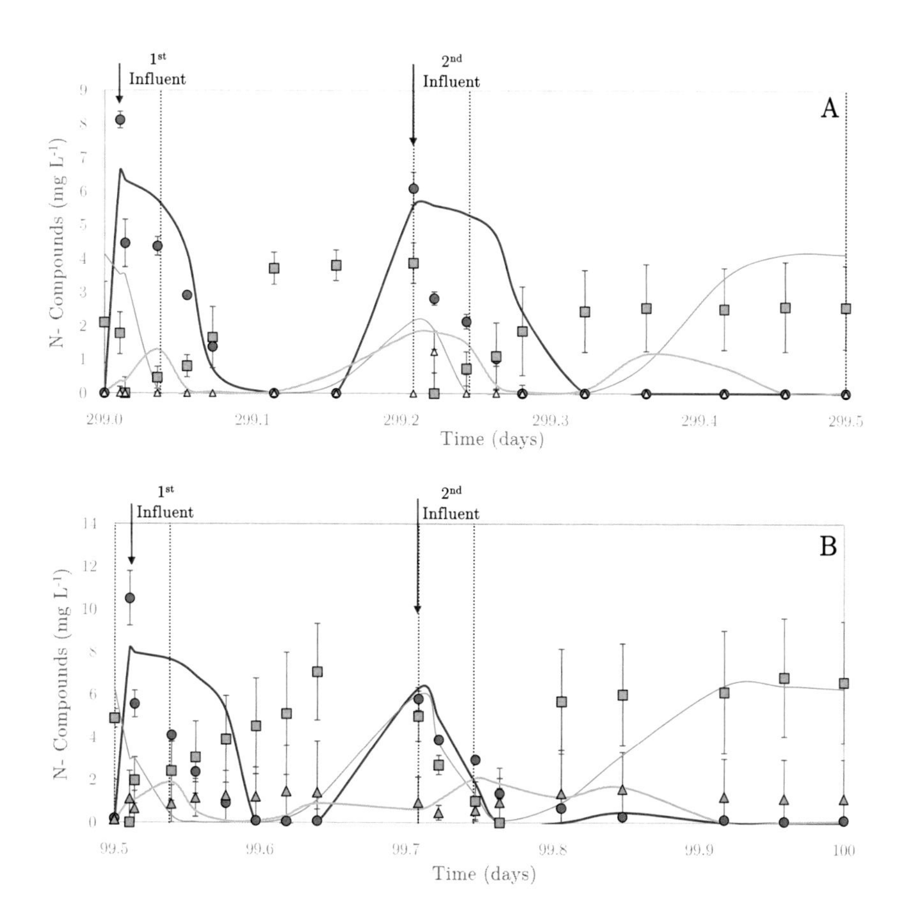

Figure 6.7. Validation of the model showing the concentrations of the N-compounds for period 1A (A) and 1C (B) with a SRT of 52 days and 17 days, respectively: NH_4^+-N (▬), NO_3^--N (▬) and NO_2^--N (▬) modelled concentrations and NH_4^+-N (●), NO_3^--N (■) and NO_2^--N (▲) measured concentration (average between of the measured cycles). The grey-shaded areas correspond to the dark phases (lights turned off) during the cycles.

The individual graphs of the modelled N-compound concentrations as well as the oxygen and COD concentrations and the measured data for the validation periods are presented in Appendix D.3. Based on the estimated IOA values, the model described satisfactorily the measured data during the three periods. Furthermore, the best descriptions were achieved for the ammonium, COD and oxygen concentrations (with IOA values closer to 1.0); meanwhile the highest errors were observed for the description of the nitrite and nitrate concentrations (with IOA values between 0.0 and 0.5).

Analysing closely the values of nitrite and nitrate shown in Figure 6.6, D.4 and D.5 (these last two in Appendix D), the nitrite concentrations were always higher than the measured ones. However, it did not exceed values higher than 2 mg NO_2^--N L^{-1}, and was always transformed into nitrate, or denitritated. With regard to nitrate, the modelled concentrations were higher than the measured concentrations. Still, looking closely at the nitrate production in each of the periods, the highest production was obtained in period 1C with a value of 6.4 mg NO_3^--L^{-1}, followed by periods 1B and 1A with values of 4.7 and 4.1 mg NO_3^--L^{-1}, respectively. Thus, following the same trend of the measured data. Period 1C had the lowest biomass concentration resulting in lower oxygen consumption from biomass respiration and less light attenuation. However, it must be noted that the nitrate production modelled is delayed in comparison with the measured data. This is associated with the availability of oxygen inside the reactor, the modelled concentration of nitrate starts when the concentration of oxygen modelled starts to increase in the bulk liquid. In fact, the measured O_2 concentrations were described closely by the model in the three periods as shown by the low errors (with IOA values close to 1). However, the concentration of oxygen also starts later than the measured data, which as explained before influenced the nitrate production.

Observing closely the production of measured oxygen concentration (Figure 6.6), it started even before ammonium was completely consumed: it started as soon as the light phase started, while this was not the case with the modelled oxygen concentration. Therefore, it could be that the oxygen production modelled was limited by the calibrated growth rate of algae, or the light extincition coefficient. Yet, the combination of calibrated parameters for period 1B resulted in lower errors, and could be further validated. Furthermore, based on the errors presented and the results obtained, the model described satisfactorily the light attenuation effect by the biomass on the oxygen production (by microalgal growth). Accordingly, as reported in Chapter 4, the highest oxygen production was observed in period 1A, followed by periods 1B and 1C, respectively. Therefore, the model was able to describe and reproduce the different oxygen production and consumption profiles occurring in the microalgal-bacterial reactor at different SRTs.

6.3.5 Growth rate in a microalgal-bacterial consortium

The maximum growth rates of the AOB, NOB, heterotrophic bacteria and algae were calibrated and validated under sequencing batch mode operation for the three experimental periods (Table 6.2). The maximum growth rates of the AOB and NOB were 1.10 (± 0.02) and 1.30 (± 0.01) d^{-1}, respectively. The maximum growth rate was the same in all three periods, hence, there was no accumulation of NO_2^--N in the reactor, neither presence of nitrite in the effluent. The differences in nitrate production in the three periods as well as the differences in ammonium consumption that can be clearly seen when comparing Figure 6.7 and Figure 6.5, are attributed to the oxygen availability in the reactor. This was calculated in Chapter 4, and explained more in detail in section 4.3.4. Therefore, the limiting step for the nitrification is the oxygen, which is ultimately related to the algal growth rate and the limiting factors affecting the photosynthesis. Therefore, the maximization of the nitrification process can be done by ensuring the presence of sufficient oxygen during the aerobic phase.

Comparing the growth rates of AOB and NOB with other studies, the growth rate of AOB is higher than the studies reported in the literature; yet, close to the value reported in Chapter 5. Furthermore, comparing in particular with the values reported in microalgal-bacterial models, Solimeno et al. (2017) observed lower values for the growth rate of AOB (0.63 d^{-1}) and NOB (1.10 d^{-1}). It must be taken into account that the experimental high rate algae pond used by Solimeno et al. (2017) for the model calibration was run in continuous mode with an HRT of 4.2 days, which could have had an implication on the retention of nitrifiers. Furthermore, in that study nitrification was performed during the night with oxygen being externally supplied during that period, hence the nitrification was not sustained by photosynthesis. Van der Steen et al. (2015) calculated, through modelling, a maximum growth rate of AOB of 0.11 d^{-1} for a microalgal-bacterial reactor with a SRT of 15 days and HRT of 1 day. These authors attributed this lower value to an overestimation of the biomass of nitrifiers, which corresponded to 18% of the total VSS in the reactor.

Finally, comparing with the previous version of the model, in Chapter 5 the growth rate was 0.90 d^{-1} and 0.65 d^{-1} for AOB and NOB, respectively; these values were not calibrated but taken from the literature (Kaelin et al., 2009). Also, the limiting step in the experiments conducted in Chapter 5 was the growth rate of NOB, contrary to the results reported in Chapter 4 and this Chapter. Thus, no nitrate was present in the N-compound profiles, and the main removal mechanism was via nitritation-denitritation (Figure 5.). Also, during the experiments conducted in Chapter 5, a real wastewater was used with a high concentration of ammonium (264 ± 10 mg L^{-1}), which might have inhibited the second step in the nitrification due to the possible inhibiton of NOB by free ammonia. In this new version of the model, the calibration of the two-step nitrification was achieved under different operational conditions, different initial concentrations, and under a continuous sequencing batch operational mode, confirming the cability of the model to describe the processes of interest under different scenarios.

Heterotrophic bacteria play an important role within the microalgal-bacterial consortia removing the organic carbon in the anoxic and aerobic phases either through denitrification or oxidation. The calibration and validation of the heterotrophic bacterial activity for the conditions tested were achieved with a growth rate of 5.5 (± 0.01) d^{-1}, which compared to other studies, is higher than the ones reported in the literature for microalgal-bacterial consortia. However, compared with activated sludge, the growth rate of heterotrophic bacteria is within the typical reported values (Table 6.2) (Metcalf & Eddy, 2002). Additionally, most of the heterotrophic bacterial biomass was active under anoxic conditions, thus the anoxic factor defined for the modelling was 0.9. Total denitrification was achieved during the anoxic phase (lights off), which lasted less than one hour. Therefore, and taking into account that heterotrophs consumed 8.66 gCOD $gNO_3^{-}\text{-}N^{-1}$ denitrified, and that at shorter SRT the nitrate production was higher (Figure 6.7), during period 1C up to 50% of the COD was removed anoxically.

Within the microalgal-bacterial consortia one of the most important processes is the growth rate of algae, and consequently the photosynthesis, especially when the main objective is to support the aerobic processes, without using any external aeration. The growth rate of algae was calibrated at 2.00 ($\pm$ 0.05) d^{-1}, while other authors have reported values for algal growth between 0.66 - 1.50 d^{-1} for algal-bacterial biomass (Solimeno et al., 2017; van der Steen et al., 2015; Zambrano et al., 2016). Comparing these values with the maximum growth rate calculated in this research, $\mu_{m,P}$ is on the high side for an algal-bacterial biomass. Moreover, algal growth rates can range from 0.1 - 11 d^{-1} (Decostere et al. 2013), such as reported by Decostere et al. (2016) with a growth rate of 0.254 d^{-1} for *Chlorella vulgaris* and Wágner et al. (2016) with a maximum growth rate for an algal biomass of 3.6 ($\pm$ 0.04) d^{-1}, calculated using the ASM-A model proposed by Wágner et al. (2016).

Furthermore, the rate of algal growth in Chapter 5 (0.85 d^{-1}) is slower than the value obtained in this chapter of 2.00 ($\pm$ 0.05) d^{-1}. These differences are attributed to the dynamics between the light extinction coefficient, biomass concentration, light intensity and water turbidity. The k was 0.0748 m^2 $gTSS^{-1}$ in Chapter 5, experimentally calculated, while in this new version of the model the k was calibrated resulting in a lower value of 0.019 m^2 $gTSS^{-1}$. This is one of the most important parameters that influence the growth of algae, especially when comparing some of the operational parameters. For instance, in Chapter 5 the light intensity was 84 ($\pm$ 3) µmol m^{-2} s^{-1}, while in this chapter (Chapter 4) it was much lower (25.9 µmol m^{-2} s^{-1}). Furthermore, the solids concentration within the reactor in Chapter 5 at an SRT of 11 days was 1.74 ($\pm$ 0.08) g TSS L^{-1}, while in period 1A (52 days SRT) the TSS was significantly higher with a value of 2.7 ($\pm$ 0.8) g TSS L^{-1}.

Therefore the light attenuation, effecting the availability of light in the reactor, is one of the most important factors for the optimization of algal growth in a microalgal-bacterial system. Similarly to this results, Solimeno et al. (2017) concluded that light attenuation in experimental ponds was one of the limiting factors for microalgal growth, reporting an increase in the growth rate of 40% to 60% when changing the light factor (which included the photoinhibition, photolimitation and light attenuation effect). Taking into account that the light extinction coefficient is a physical characteristic of the biomass, the performance of a microalgal-bacterial system could be improved by optimizing the SRT and HRT. In this regard, mathematical models can be helpful to evaluate different possible scenarios towards the optimization of microalgal-bacterial systems.

6.3.6 SRT optimization using the microalgal-bacterial model

As described previously, the calibrated and validated model can be a tool for the identification of limiting conditions and/or determination of optimal operational parameters. Therefore, the evaluation of shorter SRTs was carried out using the validated model from period 1C (17 days SRT). The SRT was shortened to values of 15, 10, 5, 3, and 1 days, in order to assess (i) the ammonium removal mechanisms, (ii) oxygen production and biomass production, and (iii) to identify a possible failure of the system at different operational conditions. The results of the N-compounds and oxygen concentrations for the SRTs of 10, 5 and 0.9 days are presented in Figure 6.8. The rest of the results at other SRT scenarios are presented in Appendix D.4.

Based on the results from the microalgal-bacterial model under the different SRTs, the optimum SRT for the microalgal-bacterial biomass lies between 5 to 10 days. This selection is based on the results of the ammonium removal rate, nitrate production, and denitrification. At the SRT of 10 days, most of the ammonium is converted to nitrate at a faster rate, and at the same time denitrification is fully achieved during the anoxic phase. At the SRT of 10 days the VSS in the reactor decreased to 1.26 (± 0.01) gVSS L^{-1}, compared to the concentration of 1.43 (± 0.01) g VSS L^{-1} at the SRT of 17 days (Table 6.1). The fractionation of biomass remained very similar between the two scenarios, during the SRT of 17 days the biomass was composed of 34.9% $\mathbf{X_P}$; 54.6% $\mathbf{X_H}$, 1.0% $\mathbf{X_{AOB}}$ and $\mathbf{X_{NOB}}$ and 9.5% $\mathbf{X_I}$. Meanwhile at the SRT of 10 days the composition was 41.7% $\mathbf{X_P}$; 50.7% $\mathbf{X_H}$, 0.7% $\mathbf{X_{AOB}}$ and $\mathbf{X_{NOB}}$,and 6.3% $\mathbf{X_I}$. Therefore, since the fractionation of the biomass as well as the influent ammonium concentration remained similar, the only difference between the two scenarios was the reduction of solids in the reactor by decreasing the SRT (having a higher wastage of solids). The light attenuation factor slightly increased from 0.32 to 0.40 when decreasing the SRT from 17 to 10 days. Therefore, oxygen production was enhanced and oxygen was not a limiting factor, because there were less solids that could decrease the light attenuation and less oxygen consumption by respiration. The ammonium removal rate for 17 days SRT was 2.12 mg NH_4^+-N L^{-1} h^{-1} (Chapter 4), while for SRTs of 10 and 5 days, the ammonium removal rate was 2.39 and 2.76 NH_4^+-N L^{-1} h^{-1}, respectively. Therefore, the reduction of the SRT helped to increase the removal rates of ammonium, due to the higher availability of oxygen at shorter SRTs. These results are similar to the findings of Chapter 5, when comparing two reactors operated with SRTs of 7 and 11 days each. In that study, the oxygen production and light attenuation were similar, but the ammonium removal rates were higher at the SRT of 7 days due to a higher oxygen availability and the lower biomass respiration.

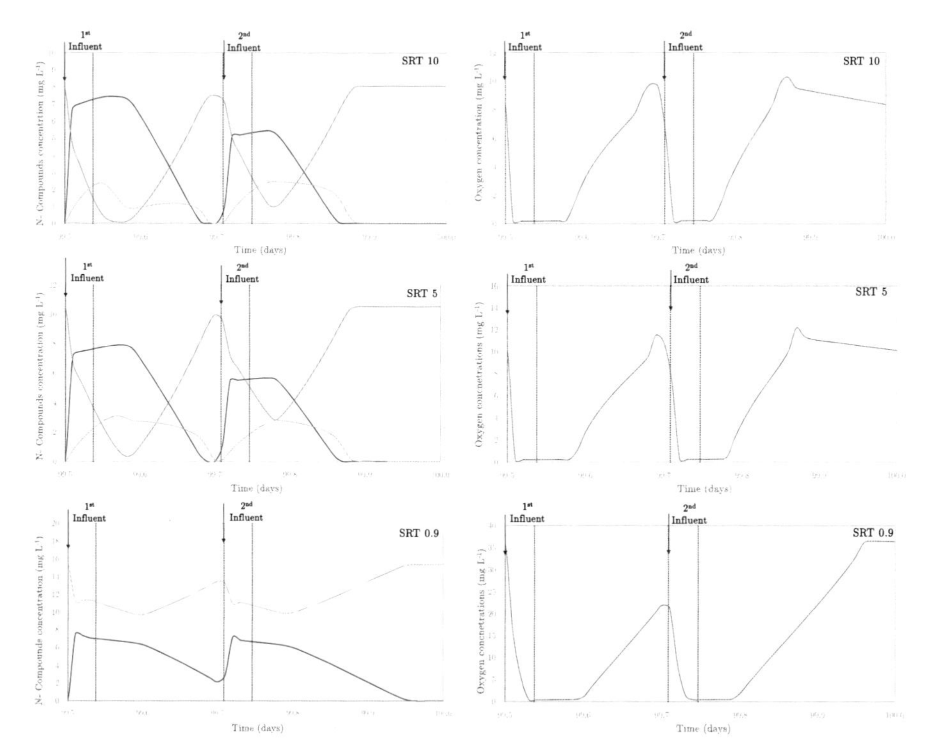

Figure 6.8. Prediction of the N-compounds and oxygen concentration for shorter SRTs: 10, 5 and 0.9 day. NH_4^+*-N (▬), NO_3^--N (▬), NO_2^--N (▬), and O_2 (▬) concentrations. The gray-shaded areas correspond to the dark phases (lights turned off) during the cycles.*

Further reduction of the SRT to values below 5 days SRT resulted in an incomplete nitrification, an increase in nitrite concentration, accumulation of nitrate, and an increase in oxygen production as seen in the peaks of Figure 6.8. The lower the

SRTs, the higher the oxygen production and accumulation once the COD and ammonium concentrations had been depleted.

The limitation of the denitrification or denitritation process could be due to the lack of organic carbon. Analysing the results of the 3 days and 1 day SRTs (Figure D.8), and taking into account the higher availability of oxygen, a hypothesis is that most of the organic carbon was removed through aerobic oxidation. In order to prove this hypothesis an extra scenario was modelled with an SRT of 3 days and increasing the organic carbon concentration to 175 mgCOD L^{-1}. The results are presented in Figure D.8 in Appendix D.4. As expected, the limitation of the denitrification was due to a lack of organic carbon, mostly because with the higher oxygen production most of the organic carbon was oxidized aerobically and became insufficient to support the denitrification or denitritation processes, resulting in a higher concentration of NO_3^--N in the effluent. Furthermore, in spite of the higher oxygen concentration, and taking into account that the reactor is operated under sequencing batch conditions, in this simulation not all the ammonium (8.0 mg NH_4^+-N L^{-1}) was fully converted to nitrate (reaching a concentration of 5.0 mg NO_3^--N L^{-1}), but also it reached a high concentration of nitrite of up to 2.4 mg NO_2^--N L^{-1}.

In any case, the applied SRT must not fall below the minimum SRT required for nitrification (SRT_{min}). The minimum SRT modelled was 1 day without showing any wash-out of the biomass. However, at 0.9 days SRT the NOB disappeared from the reactor and there was nitrite accumulation, thereafter the ammonium removal rates decreased and the denitritation stopped (Figure 6.8). The microorganisms with the lowest growth rate within the microalgal-bacterial biomass are the AOBs, which implies that these would be the first to be washed-out of the system at shorter SRTs. The SRT_{min} for AOBs, using the maximum growth rate calibrated herein, is 0.9 days, calculated using the equation proposed by Ekama and Wentzel, (2008a). Indeed, when the model was run at SRT 0.8 days, nitrification fully stopped, ammonium accumulated and the only active processes were the photosynthesis and COD aerobic oxidation. Once the model was run with an SRT of 0.5 days, after 1 day run, there was nor algal neither bacterial biomass present in the system (data non shown).

Overall, the SRT in a microalgal-bacterial reactor is the most important operational parameter, it not only determines the solids concentration in the reactor but also plays an important role in the removal mechanisms. This conclusion was also supported by the experimental data in Chapter 4. Moreover, the results of the modelled SRTs showed that for SRTs higher than 15 days, the system was oxygen limited and did not reach the highest ammonium removal rate and neither the specific removal rate required. On the opposite, at SRTs shorter than 10 days, oxygen was not limiting. However, it inhibited the denitrification process due to the faster aerobic oxidation of organic matter, becoming insufficient or unavailable in the anoxic phase for denitrification purposes. The SRT can also be used as a selective pressure for more settleable algal strains and/or faster growing AOBs. For instance, Wu et al. (2016) observed an increase in the AOBs growth rate (from 0.39 to 1.45 d^{-1}) and $K_{NH4,AOB}$ (from 0.51 to 5.23 mg N L^{-1}) at shorter SRTs on a nitrifiers biomass, and also NOB repression at a SRT of 6 days (SRTs tested from 3-15 days). In that study, the strategy was to select fast growing AOB over slow growing AOB by reducing the SRTs and at an ammonium concentration of 15 mg N L^{-1}, while ensuring the availability of sufficient dissolved oxygen.

The present study shows that mathematical models can be used as a tool to assess different scenarios to test different operational conditions, such as the SRT, pollutant concentration, and/or physical characteristics of the reactor/pond, towards the optimization of algal-bacterial systems. Further improvements to the model need to focus on the effect of phosphorous and inorganic carbon concentrations on algal growth and their effect on the different processes.

6.4 CONCLUSIONS

A microalgal-bacterial model was successfully calibrated and validated for three main operational conditions (52, 26 and 17 days SRT). The model sucessfully described the main processes of the system, namely, the two-step nitrification and denitrification, COD oxidation, algal growth and biomass production as well the biomass fractionation with regard to the different groups of microorganisms. Still, further improvements are needed in the model, related with the trend of oxygen and nitrate production in time. However, the model identified that the optimal SRT lies between 10 and 5 days. Within the optimal SRT range, the volumetric and specific ammonium removal rates were maximized, and the denitrification and COD removal processes were satisfactory. The ammonium removal rate for the SRT of 10 and 5 days was 2.56 and 2.76 mg NH_4^+-N L^{-1} h^{-1}, respectively, while it was of 2.12 mg NH_4^+-N L^{-1} h^{-1} for the 17 days SRT. The minimum SRT was defined at 0.9 days, at which the efficiency of ammonium removal starts to decrease (60%)

as well as the ammonium removal rate (0.96 mg NH_4^+-N L^{-1} h^{-1}). Furthermore, the system completely fails at 0.5 days SRT. The SRT was identified as the most important operational parameter, controlling the removal mechanisms and dynamics within the reactor. The light extinction coefficient was found to be one of the most sensitive parameters related to the physical characteristics of the biomass.

7

RESPIROMETRIC TESTS FOR MICROALGAL-BACTERIAL BIOMASS: MODELLING OF NITROGEN STORAGE BY MICROALGAE

This chapter is based on: Rada-Ariza, María, A., Lopez-Vazquez, C.M., Van der Steen, N.P., Lens, P.N.L.. Modelling of the ammonium removal mechanisms in a microalgal-bacterial sequencing-batch photobioreactor. Biotechnology and Bioengineering (Submitted)

Abstract

Respirometric tests (RT) are a common tool for assessment of microbiological processes in wastewater. Respirometric tests were used in this chapter on microalgal bacterial biomass previously cultivated in a flat panel photobioreactor. The RTs were performed successfully showing a high ammonium removal by algal uptake, reaching up to 60% of the total ammonium removed. The removal of ammonium by algae was identified to have a higher rate than nitrification mainly due to the ammonium storage capacity of the microalgae. The storage of nitrogen (ammonium) by microalgae was modelled by adapting the model presented in Chapter 6, including two new processes: (i) nitrogen algal uptake and (ii) phototrophic growth on stored nitrogen. The model was calibrated for ammonium, nitrite, nitrate and oxygen concentrations resulting in small errors (indexes of agreement exceeding 0.8). The maximum nitrogen stored was 0.3 g N_{sto} $gVSS^{-1}$ of algal biomass, while the maximum specific phototrophic growth was 3.5 and 1.2 d^{-1} for the growth on extracellular nitrogen and the growth on stored nitrogen, respectively. The maximum growth of ammonium oxidising bacteria and nitrite oxidising bacteria was 0.50 and 0.76 d^{-1}, respectively.

7.1 Introduction

The interactions between algae and bacteria are not just about the exchange of carbon dioxide and oxygen. For instance, cyanobacteria release a variety of organic molecules as presented by Abed et al. (2007), these exudates serve as carbon source for aerobic heterotrophic bacteria. While working with heterotrophic organisms from paper wastewater and cyanobacteria, Kirkwood et al. (2006) reported how the production of exudates by cyanobacteria did not completely inhibit bacterial growth, instead, they were used as a organic carbon substrate. This can be explained by natural selection since the microorganisms were already adapted to the exudates. In addition, the study reported that the exudates also enhanced the removal of dichloroacetate. Choi et al. (2010) reported the negative effect of cyanobacteria on nitrification rates, which were inhibited by a factor of four. Nevertheless, ammonium was completely removed. Other negative effects of microalgae on bacteria are the increase of pH due to the photosynthetic activity which could inhibit the growth of bacteria. Therefore, microalgal and bacterial interactions can have positive or negative effects on both microorganisms.

The interactions between microalgae and bacteria offer a large potential for bioremediation of nutrient rich wastewaters. However, some aspects need to be taken into account since they determine the removal efficiencies and the nutrient removal pathways. In order to optimize the operational parameters, which can enhance the removal of pollutants, it is necessary to understand the interactions between microalgae, bacteria, light and nutrients (Subashchandrabose et al., 2011). In order to maximize the nutrient removal efficiency, it is necessary to determine which stoichiometric and kinetic parameters are most sensitive within a microalgal-bacterial consortium. Furthermore, to analyse how these parameters can be affected by the growth and operational conditions. For instance, in Chapter 5 it was shown that the SRT is a key parameter that affects the growth rate of the microorganisms by either increasing or decreasing the solids content in the photobioreactor.

Respirometric tests are a common tool to assess the aerobic process rates and characterise the biomass (Spanjers and Vanrolleghem, 2016). A variation to this technique is the inclusion of titrimetric measurements, called respirometric-titrimetric measurements. In this variation, in addition to the oxygen profiles, information about the nitrogen removal can be determined by the dosage of acid or base in order to maintain the pH of the system (Decostere et al., 2013). Respirometry has been used for the kinetic determination of several activated sludge microorganisms (Spanjers and Vanrolleghem, 2016). In addition, Decostere et al. (2013) developed a protocol for respirometric-tritimetric measurements in algae, for the calculation of the kinetic parameters, and further calibration of a microalgae growth model.

The objective of this chapter is to apply respirometry to microalgal-bacterial biomass from a steadily performing reactor, in order to identify the most significant kinetic parameters, such as the biomass specific ammonium oxidation and the aerobic oxidation by AOB and NOB, respectively, and the total oxygen production rate. Furthermore, to get a closer look into the intracellular nitrogen storage processes performed by microalgae in a microalgal-bacterial biomass. This been widely studied in marine ecology, but there is a lack of information regarding the intracellular storage of nitrogen by microalgae in microalgal-bacterial consortia treating wastewater.

7.2 MATERIALS AND METHODS

7.2.1 Microalgal-bacterial parent reactor

A 5.75 L open flat-panel reactor (FPR) (0.25m x 0.23m x 0.1 m) with a net working volume of 4 L was operated as a sequencing batch reactor (SBR). A detailed picture of the FPR is presented in Figure 7.1. The light intensity on the reactor surface was 766.5 (± 154.1) µmol m^{-2} s^{-1}, and the temperature was controlled at about 25 °C through a heating jacket. The FPRs were completely mixed, using magnetic stirrers operated at 500 rpm. The pH in the FPRs was kept around 7.5 through the addition of a phosphate buffer solution (PBS) to the synthetic wastewater. The FPRs were operated for 331 d in cycles of 24 h with two feedings per cycle. The HRT was set to 1 day and the SRT at 10 days. The inoculation was done following the same procedure described in Chapter 3. The base reactor was fed with BG-11 medium as synthetic wastewater (Becker, 1994). The nitrogen source was ammonium and the concentration fed to the FPRs was 0.15 g L^{-1} of NH_4Cl to ensure an ammonium concentration of 40 mg NH_4^+ L^{-1} in the influent. The phosphorous concentration in the influent was 5 mg PO_4^{3-}-P L^{-1} (0.03 g L^{-1} of K_2HPO_4). The phosphate buffer used for pH control had the same concentration as defined in Chapter 3. Bicarbonate was added as a supply of inorganic carbon, at the beginning of the operation the concentration was set at 400 mg HCO_3^- L^{-1}, and from day 170 onwards the bicarbonate concentration was 700 mg HCO_3^- L^{-1} as $NaHCO_3$ (0.96 g $NaHCO_3$ L^{-1}). Acetate was added as organic carbon source with a concentration in the influent of 120 mgCOD L^{-1}. Three operational periods were defined during the operation of the reactor. The first period was between day 0 and day 130, the second period between days 130 – 170, and, finally, period 3 from day 170 to day 320. The analysis of the ammonium removal rate and biomass characterization was done following the same approach presented in Chapter 4.

Light extinction coefficient

The light extinction coefficient was calculated using the approach described in Chapter 5. The light measurements were done in 12 points along the area of incidence light and at 5 points along the depth (light path) (Figure 7.1) using a Quantum meter MQ-200 (Apogee Instruments, US). The light measurements were done at 9 different concentrations, where C1 corresponds to the actual concentration in the reactor, and C9 is the concentration of the influent medium. The dilutions from C1 to the different concentrations were done using the influent medium. The data collected from this experiment were used to determine the extinction coefficient, k, in Eq. 5.2 using the MS Excel tool *Solver* (GRG nonlinear algorithm).

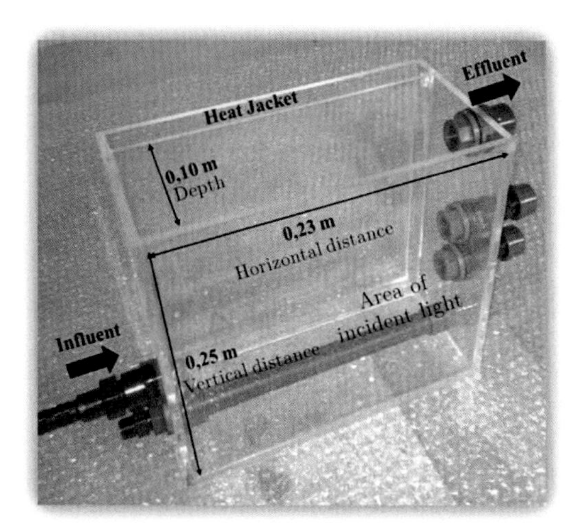

Figure 7.1. Flat panel reactor used as parent microalgal-bacterial reactor, the light was applied perpendicular to the largest cross-sectional area.

7.2.2 Respirometric test methodology

Once the nitrification process reached steady conditions in the parent microalgal-bacterial reactor, the respirometry tests were performed using the enriched microalgal-bacterial biomass. The respirometry unit that was used for these tests can be seen in Figure 7.2. The unit consists of a double-heat jacketed reactor of 1 L of volume connected to a 10-mL double heated jacketed respirometer vessel, in which oxygen was measured online using an oxygen probe WTW Oxy 3310 electrode (Weilheim, Germany). The reactor and the respirometric vessel have both a magnetic stirrer that ensures complete mixing of the biomass. The respirometric unit was illuminated (respirometric vessel and reactor) using LED lights (Phillips, The Netherlands) with an average light intensity of 310 (± 52.8) µmol m^{-2} s^{-1}. The microalgal-bacterial biomass was placed in the reactor, and additions of different compounds were performed in this reactor to evaluate the different algal and bacterial processes. Biomass was pumped from the reactor to the respirometer vessel for 30 seconds, while online oxygen measurements were recorded. Samples for the analysis of different compounds of interest were taken simultaneously from the reactor. The reactor of the respirometer unit was air tight, and it was connected to a vessel that contained a sodium hydroxide solution with the aim of entrapping the carbon dioxide produced.

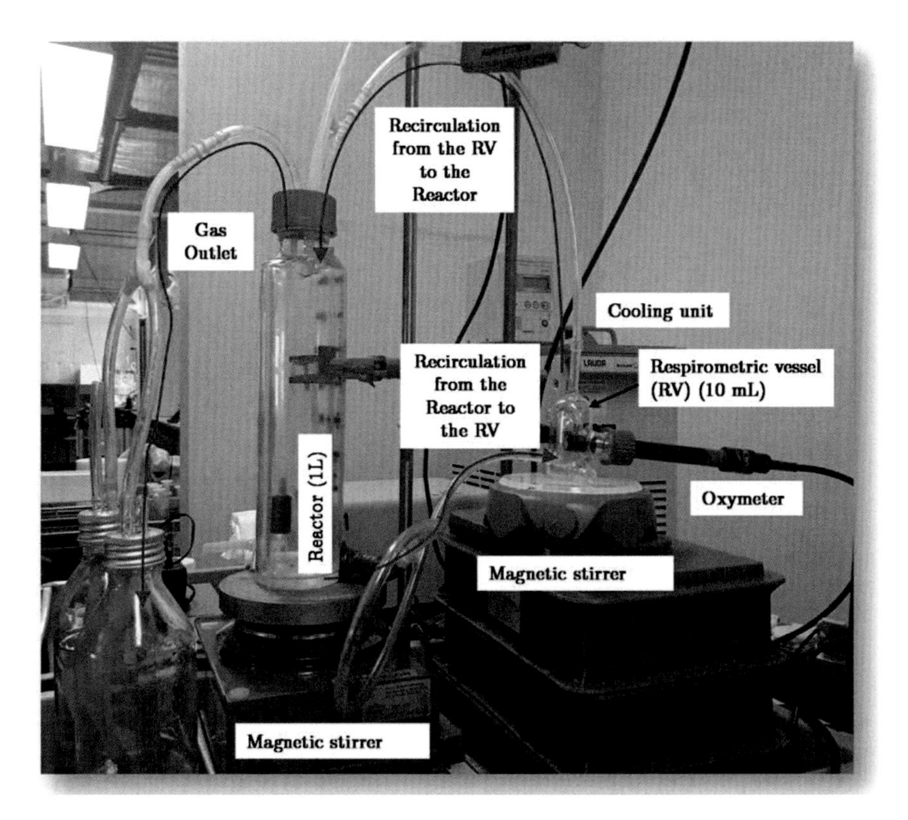

Figure 7.2. Respirometric unit used to perform the respirometric tests. Reactor of 1 L connected through a pump (not in picture) to the double wall heated respirometer vessel (RV).

General steps to perform the respirometric tests

1) Prior to the conduction of the RTs, light measurements were performed over the respirometric unit to ensure an approximate incident light of 300 µmol m^{-2} s^{-1}.
2) 0.5 L of biomass were withdrawn from the microalgal-bacterial base reactor at the end of the cycle before the settling time started. Using the withdrawn volume and prior to the start of the test, samples were collected for the determination of the following parameters: ammonium, phosphate, VSS, TSS and alkalinity.
3) The volume used for the respirometric tests was 0.4 L, this volume was placed in dark conditions (by covering the respirometric unit) in the reactor of the respirometric unit. Thereafter, N_2 was flushed in order to remove the dissolved O_2 from the sample.
4) After the placement of the biomass and when the dissolved O_2 concentration dropped below the detection limit, the medium was added. This medium contained all the micronutrients and macronutrients of the modified BG-11 medium fed to the base reactor. The only variation for certain tests were the concentrations of ammonium and inorganic carbon (bicarbonate).
5) Depending on the objective of the test, allylthiourea (ATU) was added to inhibit nitrification (like in tests RT-2 and RT-3).
6) After the addition of the medium, and it was completely mixed (for 20 to 30 seconds), biomass was recirculated from the respirometric reactor to the respirometric vessel, and the light was turned on to promote photosynthesis. In order to avoid any oxygen saturation in the respirometric vessel, once the

concentration reached values between 7 – 8 mgO_2 L^{-1}, the light was turned off, and the respirometric reactor was covered with an aluminium foil. The dark conditions promoted the respiration of the biomass and further decrease of the O_2 concentration. At the same time, recording the oxygen consumption allowed to calculate the biomass respiration rate. Alternatively, when nitrification occurred, the consumption of O_2 by this aerobic process was determined.

7) Once the oxygen concentration reached in between 1 – 2 mg O_2 L^{-1}, the algae-bacteria mixture was recirculated from the respirometric vessel to the respirometric reactor and viceversa using an external pump to ensure similar conditions in the respirometric reactor and respirometric vessel (e.g. same nitrogen and inorganic carbon concentrations).
8) Samples were collected in the respirometric reactor. At the beginning of the tests, the sampling was more frequent but less frequent towards the end. Usually, measurements were taken at the beginning every 5 to 15 minutes, and towards the end, every 30 to 60 minutes.
9) The RTs had usually a duration between 5 to 8 hours due to practical limitations, or until there was no more oxygen production.
10) At the end of the tests the following analysis were performed using the remaining biomass: ammonium, phosphate, VSS, TSS and alkalinity.
11) All analytical parameters were determined in accordance to Standard Methods (APHA, 2005), and following the methodology described in Chapters 3 and 4.

7.2.3 Modelling of nitrogen storage and utilization of stored nitrogen by microalgae

The modelling of the respirometric tests was done using the microalgal-bacterial model described in Chapter 6. The only two new processes included were the ammonium storage by microalgae, and the growth on stored nitrogen (Appendix E). Furthermore, the model was run in batch mode for each respirometric test. The initial values of the biomass characterization were calculated according to the methodology described in Appendix B.

Storage of nitrogen by microalgae

The storage of nitrogen by algae has been documented by other authors (Fong et al., 1994; Mooij et al., 2015; Wágner et al., 2016); however, to the best of our knowledge, it is not yet included in any of the mathematical models developed for microalgal-bacterial systems. In this study, the model of the nitrogen storage process by algae was conceptualized following the approach proposed by Sin et al. (2005) in which the storage and growth processes occur simultaneously. Therefore, the total nitrogen uptake by algae can be either stored within the cell and the remaining can be used for biomass growth.

The rate and maximum capacity of nitrogen stored by algal biomass was defined following the equations proposed by Sin et al. (2005) and Wágner et al. (2016), and using a Monod-type function for its description. The nitrogen storage rate depends on the ammonium concentration in the medium, on the maximum intracellularly N storage capacity of the algal cell, and the minimum nitrogen required for maintenance (f_n). Mooij et al. (2015) reported the uptake of ammonium nitrogen by algae in dark conditions, suggesting that the availability of light does not condition this process. The following kinetic equation was proposed for modelling the storage products in algae:

$$\frac{df_N}{dt} = \frac{S_{NH_4}}{S_{NH_4} + K_{S,NH_4}} k_{sto,N} \frac{f_{n,max} - f_n}{(f_{n,max} - f_n) + K_{STO,N}} X_P$$

Where:

$\frac{df_N}{dt}$: Nitrogen storage rate (gN m^{-3} d^{-1})

X_N: Concentration of stored nitrogen in the algal biomass (gN m^{-3}).

S_{NH_4}: Concentration of ammonium nitrogen (gN m^{-3}).

K_{S,NH_4}: Microalgal saturation constant for growth on ammonium nitrogen (gN m^{-3}).

$k_{sto,N}$: Storage rate of nitrogen in the algal biomass (gN gCOD$_{X_P}^{-1}$ d^{-1}).

$f_{n,max}$: Maximum fraction of nitrogen stored in the microalgal biomass, the fraction is expressed in grams of nitrogen per gram of microalgal biomass (X_P) (gN gCOD$_{X_P}^{-1}$).

f_n: Minimum fraction of nitrogen stored in the microalgal biomass, the fraction is expressed in grams of nitrogen per gram of microalgal biomass (X_P) (gN gCOD$_{X_P}^{-1}$).

$K_{STO,N}$: Saturation constant for nitrogen storate (gN gCOD$_{X_P}^{-1}$)

X_P: Microalgal biomass (gCOD$_{X_P}$ m^{-3}).

For this process the parameters selected for calibration were: $k_{sto,N}$, $f_{n,max}$, and $K_{STO,N}$. These values were selected for calibration due to the lack of information related to the storage of nitrogen in microalgal-bacterial consortia.

Growth of microalgae on nitrogen stored by microalgae

The growth of microalgae on stored nitrogen was based on the equation proposed by Sin et al. (2005). The growth of algae on stored nitrogen would depend on the ammonium concentration in the medium and the effect of the light attenuation. The mathematical expression that describes the use of the stored nitrogen is composed by two parts. The first part is expressed as a Monod function, in which the use of stored nitrogen depends on the half-saturation constant for growth on X_N. The second term regulates the use of stored nitrogen based on a regulation constant of the cell ($f_{X_N}^{REG}$). Then, when $\frac{X_N}{X_P}$ is higher then the use of the stored nitrogen will be high depending on the regulation constant. It is assumed that the rate of utilization of stored nitrogen is different from the growth rate on external nitrogen substrate. The kinetic process proposed for the utilization of the stored nitrogen (X_N) is:

$$\frac{dX_P}{dt} = \mu_{m,STO,N} \frac{S_{NH_4}}{K_{NH_4,P} + S_{NH_4}} \left\{1 - \exp\left(\frac{-I_o[1 - \exp(-k\, X_T\, L)]}{k\, X_T\, L\, I_s}\right)\right\} \left\{\frac{\frac{X_N}{X_P}}{K_{STO,N,P} + \frac{X_N}{X_P}} \frac{\frac{X_N}{X_P}}{f_{X_N}^{REG}}\right\} X_P$$

Where:

$\mu_{STO,N}$: Microalgal maximum growth rate on stored nitrogen (d^{-1}).

$K_{STO,N,P}$: Half-saturation constant for growth on X_N (g N_{STO} g COD_{XP}^{-1}).

$f_{X_N}^{REG}$: Regulation constant of the microalgal biomass controlling the growth of microalgal on X_N (g N_{STO} g COD_{XP}^{-1}).

For this process the parameters selected for calibration were: $\mu_{STO,N}$, $K_{STO,N,P}$, and $f_{X_N}^{REG}$. These values were selected for calibration due to the lack of information related to the storage of nitrogen in microalgal-bacterial consortia.

7.3 RESULTS AND DISCUSSION

7.3.1 Solids concentration and light attenuation coefficient in the base microalgal-bacterial reactor

The TSS concentration in the parent reactor (Figure 7.3) was 1.78 (± 0.22), 2.13 (± 0.19) and 1.56 (± 0.18) gTSS L^{-1} for periods 1, 2 and 3, respectively. The concentration of VSS was 1.43 (± 0.19), 1.82 (± 0.09) and 1.27 (± 0.17) gVSS L^{-1} for periods 1, 2 and 3, respectively. The highest biomass concentration was measured in period 2, while in period 3 the biomass reached steady-state conditions. The standard deviation was not higher than 11% compared to the average concentration during period 3. The solids in the effluent along the three periods were not higher than the 15% of the solids in the reactor, which suggests that there was a good biomass retention within the reactor.

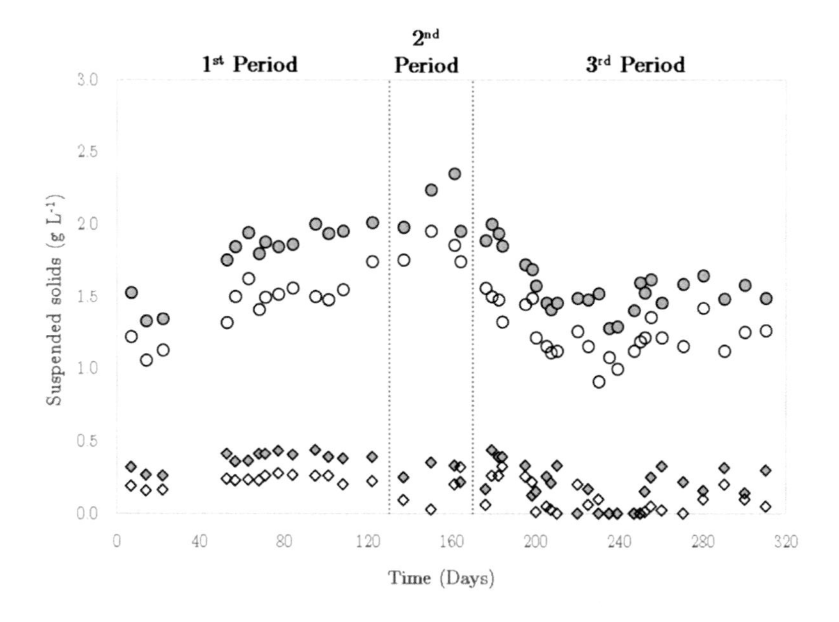

Figure 7.3. Suspended solids concentrations during the entire operation of the microalgal-bacterial reactor. TSS in the reactor (●), VSS in the reactor (○), effluent TSS (◆), and effluent VSS (◇).

The light extinction coefficient of the microalgal-bacterial biomass in the reactor was calculated using the light intensities measured with a submerged light meter. These measurements were done along the surface area of the incident light in 4 points (horizontal distance), in 3 points along the height of the reactor (vertical distance, and 5 different points (depth) along the light path of the reactor (total depth of the reactor 10 cm). Figure 7.4 shows that at concentration 1 (concentration in the reactor), the light intensity is zero after 3.5 cm. Solving Equation *(5.2)* using the measurements herein, the light extinction coefficient for this microalgal-bacterial biomass was calculated to be around 0.0763 ($\pm$ 0.0075), m^2 $gTSS^{-1}$. This value was used for the calibration of the model.

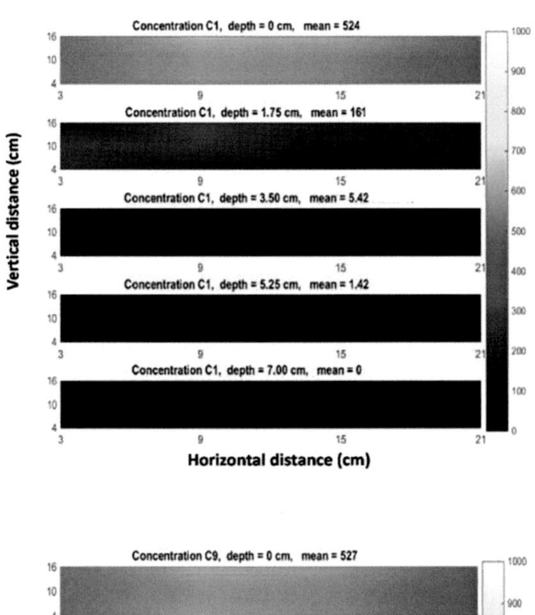

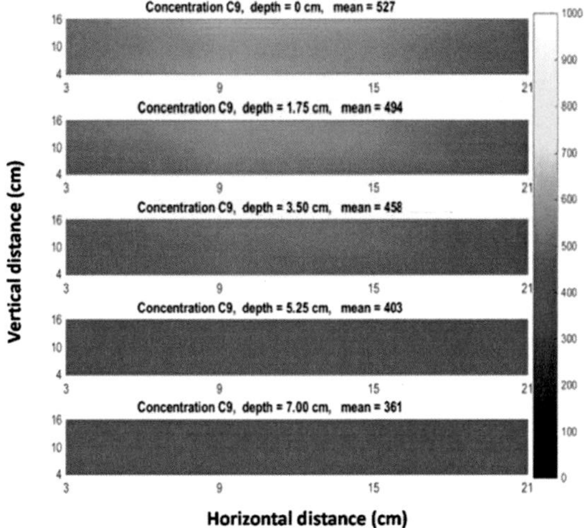

Figure 7.4. Light measurements (µmol m^{-2} s^{-1}) in the flat panel reactor at C1: 1.56 (± 0.18) gTSS L^{-1} and C9 corresponds to zero as it corresponds to the synthetic medium fed to the reactors.

7.3.2 Ammonium removal rates, efficiency and biomass characterization of the base microalgal-bacterial reactor

The parent reactor used to cultivate the microalgal-bacterial biomass for later use in the respirometric tests was operated steadily for 310 days, and the entire operation was divided in three periods. During the three periods (

Figure 7.5), the ammonium removal efficiency was 94.7 (± 4.0) %. The ammonium removal rate (ARR) between the three periods was not significantly different ($p>0.05$), with values of 3.26 (± 0.30), 3.34 (± 0.51) and 3.21 (± 0.24) mg NH_4^+-N L^{-1} h^{-1} for period 1, 2 and 3, respectively. The ammonium removal rate of algal biomass and bacterial biomass was calculated for periods 1 and 3, whereas for period 2 this was not possible. During period 2 possibly simultaneous nitrification/denitrification occurred. Therefore, it was difficult to differentiate between how much ammonium was removed by algae and how much by nitrifiers. The ammonium removal rate by nitrifiers was 2.49 (± 0.46) and 1.87 (± 0.32) mg NH_4^+-N L^{-1} h^{-1} for period 1 and 3, respectively. On the other hand, the ammonium removal rate by algae was lower than by nitrifiers with values of 0.77 (± 0.63) and 1.34 (± 0.38) mg NH_4^+-N L^{-1} h^{-1} for periods 1 and 3, respectively.

The simultaneous nitrification/denitrification process that took place during period 2 is probably due to the presence of anoxic conditions within the reactor, e.g. the O_2 concentration droped below detection limits soon after the influent feeding. The decrease in oxygen concentration during this period could be caused by the increase in biomass towards the last days of period 1 and the start of period 2 (Figure 7.3), as well as by the higher activity of the aerobic processes and biomass respiration.

Therefore, during period 3 the inorganic carbon was increased from 0.4 (period 1 and 2) to 0.7 g HCO_3^- L^{-1}. This led to an increase in algal activity that increased the oxygen generation and avoided the development of anoxic conditions. Furthermore, it increased the contribution of algae to the removal of ammonium. Since the main objective was to have a nitrifying biomass under steady-state conditions to be used in the respirometric tests, in order to confirm that denitrification did not take place, an evaluation of one of the cycles in period 3 was carried out (data non shown). The N-compound concentrations were measured every half an hour during the entire reaction time of the sequencing batch operation, showing the absence of denitrification (e.g. nitrate produced was not removed), and the oxygen concentrations never decreased below 6 mg O_2 L^{-1}.

The total specific ammonium removal rate of the system was 0.05, 0.04 and 0.06 gNH_4^+-N gVSS^{-1} d^{-1}. Therefore, the highest biomass activity was observed in period 3, and during this period all the respirometric tests were performed. Also in this period, the biomass was characterized. The total biomass was comprised of 80.6 ($\pm$ 10.8) % microalgae, 17.8 ($\pm$ 9.9) % heterotrophic bacteria, 1.2 ($\pm$ 0.6) % ammonium oxidizing and 0.5 ($\pm$ 0.2) % nitrite oxidizing bacteria.

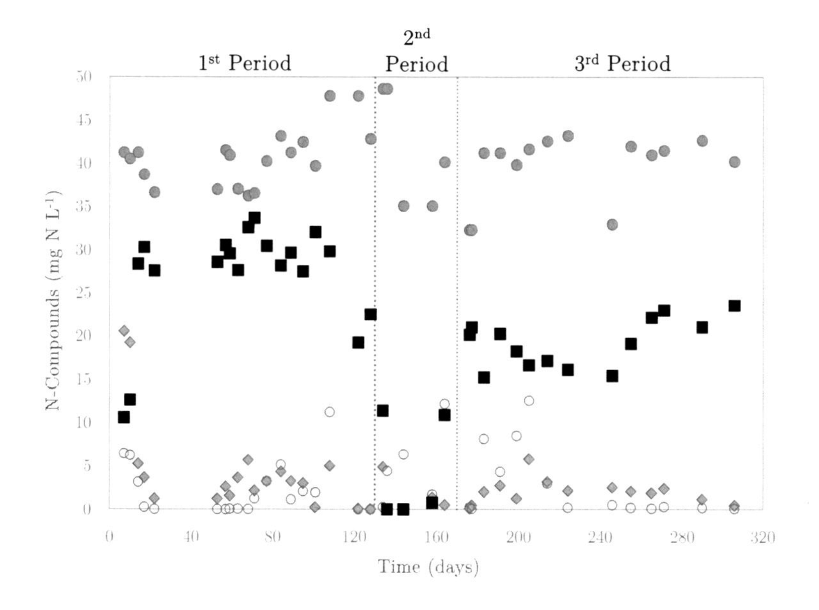

Figure 7.5. Nitrogen compounds concentrations during the entire operation of the microalgal-bacterial reactor. Influent NH_4^+-N (●), effluent NO_2^--N (◇), effluent NO_3^--N (■), and effluent NH_4^+-N (○).

7.3.3 Nitrogen storage by microalgae in a microalgal-bacterial biomass

Respirometric tests were conducted according to the methodology described in the materials and methods and using the microalgal-bacterial biomass cultivated in the base reactor during the third period. Figure 7.6 presents the result of a respirometric test performed on day 175. The initial concentrations were 13.4 mg NH_4^+-N L^{-1} for ammonium, 11.6 mg NO_2^--N L^{-1} for nitrite and 10.3 mg NO_3^--N L^{-1} for nitrate. As seen in the respirometric test-1 (RT-1) (Figure 7.6), during the first 13 minutes, there is a rapid decrease in ammonium concentration at a rate of 45 mg NH_4^+-N L^{-1} h^{-1}. However, this rapid decrease does not match with the production of nitrate and neither with nitrite. Furthermore, after the rapid decrease of ammonium stopped, the ammonium removal rate (0.51 mg NH_4^+-N L^{-1} h^{-1}) matched with the rates of nitrite and nitrate (0.25 (nitrite) + 0.34 (nitrate) = 0.59 mg NH_4^+-N L^{-1} h^{-1}) (between 0.6-9.2 hours).

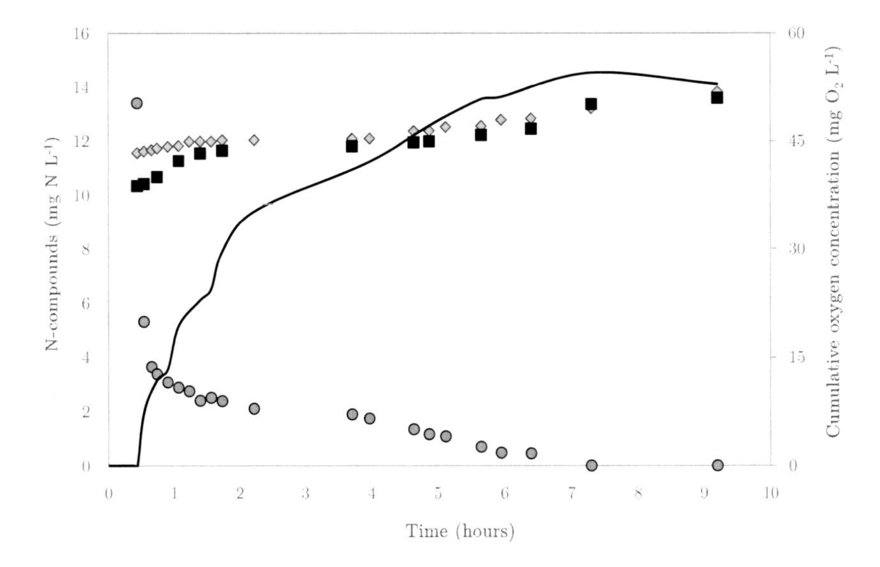

Figure 7.6. RT-1 with initial ammonium concentration of 13.4 mg NH_4^+-N L^{-1}. NH_4^+-N (●), NO_2^--N (◇), NO_3^--N (■), and O_2 (—).

The ammonium removal efficiency reached 100%, from which 42% (5.51 mgNH_4^+-N L^{-1}) was removed through nitritation and nitrification and 58% (7.84 mgNH_4^+-N L^{-1}) by algal uptake. Moreover, other possible explanations for such a rapid decrease can be ruled out. Ammonium volatilization was not a potential removal mechanism since the pH was maintained at 7.5. Another possible removal pathway is through the adsorption of ammonium into the biomass; however, there is not literature reporting such a high adsorption rate in microalgal biomass. Also, no organic carbon was present during the RT, therefore simultaneous nitrification/denitrification could not take place. In addition, controlled respirometric tests using the effluent from the reactor (without biomass) were executed to rule out any chemical precipitation as an ammonium removal mechanism (data not shown).

The total net oxygen production rate during the entire test was 0.0202 gO_2 L^{-1} h^{-1} (taking into account solely the light phase), and since the respirometer vessel was air tight, the oxygen production was entirely provided by photosynthesis. The ammonium removal rate was 1.5 mg$NH_4^+ - N$ L^{-1} h^{-1} for the microalgal-bacterial biomass. The specific ammonium removal rate of the system was 0.98 g$NH_4^+ - N$ $gVSS^{-1}$ h^{-1}, and the specific ammonium removal rates of AOB and NOB were estimated around 0.009 g$NH_4^+ - N$ $gVSS^{-1}$ h^{-1} and 0.031 g$NH_4^+ - N$ $gVSS^{-1}$ h^{-1}, respectively. The total ammonium removal through nitrification/nitritation was 5.5 mg NH_4^+-N L^{-1}, while algae removed 7.8 mg NH_4^+-N L^{-1}.

The rapid and high uptake of ammonium by algae is considered as a luxury uptake. Comparing the total production of oxygen with the total ammonium consumed by algae in this test, it can be observed that not all the ammonium was utilized for growth. Thus, it was assumed that it was likely stored intracellularly in the form of inorganic nitrogen pools (Lavín and Lourenço, 2005). The total oxygen produced was 0.0820 g O_2 L^{-1} (including the net oxygen, biomass respiration and consumption by nitrifiers), calculated during the duration of the RT (9.5 hours). The total amount of ammonium required to produce this oxygen concentration can be calculated using the yield of oxygen on ammonium proposed by Mara (2004) of 16.85 g NH_4^+-N $g^{-1}O_2$. Therefore, the ammonium required to produce the measured O_2 concentration is 4.86 mg NH_4^+-N L^{-1} and, since the total amount of ammonium removed by the algae in the 9.5 hours was 7.84 mg NH_4^+-N L^{-1}, it leaves 2.97 mg NH_4^+-N L^{-1} of ammonium to be stored in the algal cells.

In order to assess the nitrogen uptake by algae in a microalgal-bacterial biomass, a nitrification inhibitor (allylthiourea) was added during the respirometric test-2 (RT-2) (Figure 7.7). Figure 7.7 shows the result of this test, which was divided in two parts, in the first part there is ammonium removal after a pulse addition of medium (containing all nutrients except organic carbon), and after the 5th hour the oxygen production stopped. Therefore, only ammonium was supplied, as it was considered to be the limiting step. From this time onwards, the second part of the experiment starts, during which ammonium was rapidily removed within the first half an hour. However, after this second ammonium addition, the production of oxygen did not increase despite that ammonium was not the limiting (still 8.64 NH_4^+-N L^{-1} remaining). Thus, inorganic carbon was added. Subsequently, after the re-addition of inorganic carbon, the ammonium removal rate was similar to the one observed in the first part of the experiment. The total ammonium removed was 2.5 and 10.6 mg$NH_4^+ - N$ L^{-1} for the first and second part, respectively.

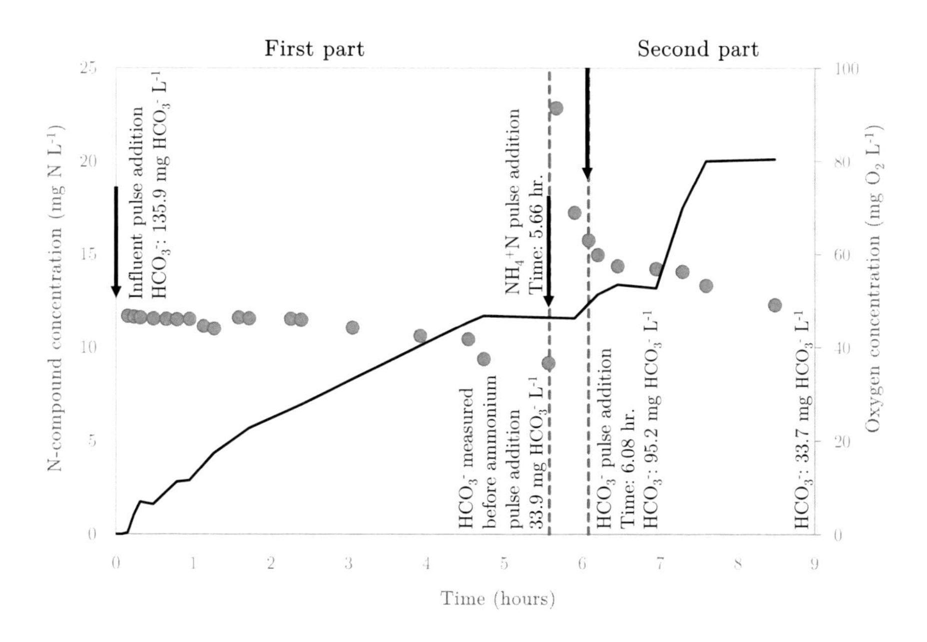

Figure 7.7. RT-2 with initial ammonium concentration of 13.4 mg NH_4^+-N L^{-1} and addition of ATU to stop nitrification and nitritation. NH_4^+-N L^{-1} concentration (●) and O_2 concentration (▬).

In the first part the total oxygen production rate was 0.0306 gO_2 L^{-1} h^{-1} and the total ammonium removal rate was 0.46 mg$NH_4^+ - N$ L^{-1} h^{-1}. The total specific ammonium removal rate was 0.391 g$NH_4^+ - N$ $gVSS^{-1}$ h^{-1}. The total inorganic carbon consumed during this part was 0.10 $gHCO_3^-$ L^{-1}.

In the second part, the total oxygen production rate was 0.0228 gO_2 L^{-1} h^{-1}, and as seen in Figure 7.7, two ammonium removal rates can be identified. A maximum removal rate of 14.76 mg$NH_4^+ - N$ L^{-1} h^{-1} within the first half an hour, and 1.01 mg$NH_4^+ - N$ L^{-1} h^{-1} after the rapid decrease on ammonium. The total specific ammonium removal rate was 3.14 $gNH_4^+ - N$ gVSS^{-1} h^{-1} and the total amount of inorganic consumed during this part was 0.06 gHCO_3^- L^{-1}.

An approximate determination of the ammonium used for growth and stored within the cell was done linking the oxygen production with the total uptake of ammonium. The total oxygen produced during the first part of the respirometric test is 0.0577 gO_2 L^{-1}, which would require the uptake of 3.42 mg NH_4^+-N L^{-1} to support the microalgal growth (16.85 gO_2 gNH_4^+-N^{-1}), assuming that biomass growth occurred only in the light phase, hence when oxygen was produced. Yet, comparing this value with the uptake of ammonium calculated from the measurements of 2.52 mg NH_4^+-N L^{-1}, it is concluded that the total amount of ammonium taken up is not enough, and that 0.9 mg NH_4^+-N L^{-1} extra are required to produce the amount of oxygen measured during the first part. However, it must be taken into account that this amount is very low and probably statistically not significant. These results confirm that the mass balances for oxygen and ammonium can be linked using the quoted stoichiometry. In the second part, the total oxygen production was 0.0401 gO_2 L^{-1}, and the total ammonium consumed by algae was 10.56 mg NH_4^+-N L^{-1}. Applying the same approach used in part 1, the nitrogen necessary to produce the total amount of O_2 is 2.38 mg NH_4^+-N L^{-1}, which indicates that 8.17 mg NH_4^+-N L^{-1} was more likely to be stored inside the cell, as there was no ammonium volatilization (pH controlled), and any chemical precipitation was ruled out.

In microalgal-bacterial systems, algae can be exposed to nitrogen limiting conditions, as reported in Chapters 4 and 5, in which most of the ammonium was removed through nitrification. Also, in Figure 4.4 (Chapter 4), it can be observed that even after ammonium was depleted, there is oxygen production by microalgae. These observations support the potential intracellular storage of nitrogen since algae might have produced the oxygen using the internal ammonium stored in the cell. However, strictly, this was not observed during the RT-2 (Figure 7.7), as towards the end of the 5th hour (last hour of the first part) there was still ammonium present in the medium. Yet, 0.9 mg NH_4^+-N L^{-1} extra was necessary in order to produce the total oxygen measured in the first part. When reviewing the inorganic carbon concentrations measured in both parts for the RT-2, the average bicarbonate concentration at the end of both parts (at which the oxygen production stopped) is 33.82 ($\pm$ 0.14) mg HCO_3^- L^{-1}. Therefore, it can be inferred that the system could have been limited by the low inorganic carbon concentration. Furthermore, to support this hypothesis, it can be stated that the inorganic carbon is necessary for both nitrification and algal uptake, therefore being also a limiting nutrient in microalgal-bacterial systems.

7.3.4 Phototrophic growth on stored nitrogen

In order to assess the growth of microalgae on the nitrogen stored, the microalgal-bacterial biomass was washed prior to the RT to ensure that neither inorganic carbon nor inorganic nitrogen was present in the mixture. Subsequently, it was placed in the respirometric reactor and a pulse addition of medium without ammonium was supplied. The results of the respirometric test-3 (RT-3) are presented in Figure 7.8.

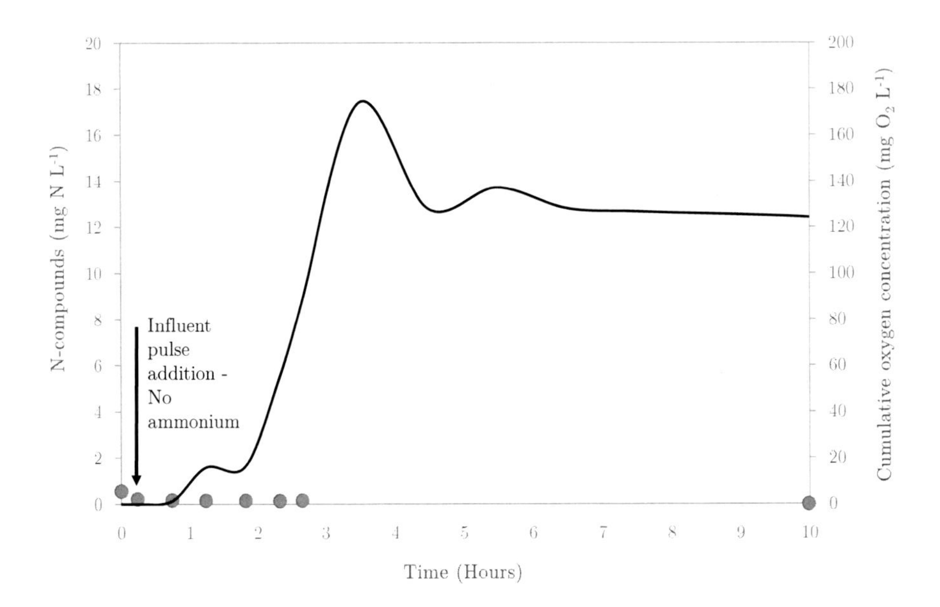

Figure 7.8. RT-3 with initial ammonium concentration of 0.22 mg NH_4^+-N L^{-1} and addition of ATU to stop nitrification and nitritation. NH_4^+-N L^{-1}. NH_4^+-N (●), NO_2^--N (◆), NO_3^--N (■), and O_2 (—).

As seen in the respirometric test-3 (RT-3), despite that the concentrations of ammonium in the medium were below 0.5 mg NH_4^+-N L^{-1}, the oxygen concentration reached a maximum concentration of 174 mg O_2 L^{-1}, and a maximum production rate of 0.13 gO_2 L^{-1} h^{-1}. In order to produce such a concentration of oxygen, the ammonium required is 10.3 mg NH_4-N L^{-1}. The concentration of HCO_3^- at the beginning of the test was 126.6 mg HCO_3^- L^{-1} (soon after addition), and at the end of the test the bicarbonate concentration was 20.6 mg HCO_3^- L^{-1}. Since bicarbonate at the end of this test was still available, it can be assumed that the oxygen production was only limited by the concentration of nitrogen and, being absent in the liquid phase, the only potential nitrogen source available was the intracellularly-stored nitrogen. These results support nitrogen storage is an algal mechanism use to remove N and utilize it in subsequent N-deprived periods.

Overall, the nitrogen stored measured and calculated in the respirometric tests presented herein was 2.97 mg NH_4^+-N L^{-1} for RT-1, and 0.9 and 2.87 mg NH_4^+-N L^{-1} for the parts 1 and 2 of RT-2, respectively, and finally 10.3 mg NH_4-N L^{-1} in RT-3. Expressing these values in terms of nitrogen stored per gram of algal biomass, the values obtained are 0.0026 gN $gVSSx_p^{-1}$ for RT-1, 0.0007 and 0.0068 gN g $VSSx_p^{-1}$ for RT-2, and 0.011 gN $gVSSx_p^{-1}$ for the RT-3. Wágner et al. (2016) reported maximum values of nitrogen stored for algal biomass for different experiments of 0.012 ($\pm$ 0.003) gN $gCOD^{-1}$ and the minimum stored value reported was 0.009 ($\pm$ 0.004) gN $gCOD^{-1}$. Quinn et al. (2011) reported a maximum value of up to 15% of nitrogen per gram of biomass. The values presented herein are lower. However, the values reported in previous studies have been obtained in enriched algal cultures, whereas in this study they have been observed in a mixed algal-bacterial culure, which could have played a role in the intracellular storage processes.

Ammonium is the preferred nitrogen compound by algae among the three different inorganic nitrogen concentrations usually available in natural or wastewater flows (Hellebust and Ahmad, 1989; Lavín and Lourenço, 2005). This inorganic compound is taken up by algae and assimilated either by the glutamine cycle or via the metabolic pathway of glutamate dehydrogenase (Hellebust and Ahmad, 1989). The last metabolic pathway has been reported for some species such as *Chlorella* under high ammonium concentrations (Hellebust and Ahmad, 1989). Furthermore, Mooij et al. (2015) reported the storage of ammonium nitrogen under dark conditions. The maximum nitrogen stored during the dark phase was 16.6 mg N L^{-1}. The storage of nitrogen described and measured by Mooij et al. (2015) occurred in a N-limited environment by uncoupling the carbon fixation (light phase) from the ammonium uptake (dark phase). In the light, all nutrients and inorganic carbon were fed with the exception of ammonium, which was added at the beginning of the dark phase. Other conditions for nitrogen storage have been reported by Lavín and Lourenço (2005) when comparing the nitrogen storage under inorganic carbon limited and non-limited conditions. The results showed that there was a high accumulation of inorganic nitrogen in both scenarios during the first days of the culture. Also, the concentration of stored ammonium decreased when the nitrogen in the medium was limiting, but inorganic carbon was sufficient (Lavín and Lourenço, 2005). Finally, they concluded that the availability of inorganic carbon influences the accumulation of inorganic nitrogen, in some scenarios the high N found in the inorganic nitrogen pools in algae was effected by the limitation of inorganic carbon. In this study, similar observations were obtained indicating that a mixed algae-bacteria culture can also have an intracellular nitrogen storage process for its further utilization as a source of nitrogen for growth or oxygen production.

7.3.5 Modelling the nitrogen storage by algae in a microalgal-bacterial biomass

The modelling of the nitrogen storage processes was done by including two new processes in the model proposed in Chapter 6: (i) nitrogen storage and (ii) growth of microalgae on stored nitrogen. The proposal to include these two processes was made following a combined approach of two different models proposed by Sin et al.

(2005) and Wágner et al. (2016). The first one presents a new approach for modelling of simultaneous growth and storage of organic carbon, while the second presents a biokinetic model for algae based on the activated sludge models, and taking into account nitrogen storage by algae. The model considers that the growth of algae on external and internal nitrogen as well as the storage of nitrogen occur simultaneously. Therefore, part of the nitrogen uptake by algae was stored as X_N, and the rest was used for growth. Furthermore, the use of X_N for algae growth was assumed to occur at a lower rate than the growth on extracellular nitrogen represented by $\mu_{P,STO}$.

Respirometric tests used for calibration

Three respirometric tests were selected for calibration. The three tests selected exhibit a fast decrease in ammonium, and based on the oxygen concentrations and ammonium consumed, it was concluded that ammonium was stored intracellularly. Data from the respirometric test-4 (RT-4) (Figure 7.9) was the first set used for calibration. For this test the biomass was exposed to light conditions for 2.5 hours in order to exhaust some of the ammonium and/or inorganic carbon left from the cycle in the base reactor. After that, a pulse of medium addition was supplied to ensure an ammonium concentration of 22.5 mg NH_4^+-N L^{-1}. The total ammonium removed in Figure 7.9 was 19.6 mg$NH_4^+ - N$ L^{-1}, from which 40% was removed through nitritation and nitrification and 60% by algal uptake. This 60% of removal of ammonium includes the storage of nitrogen within the first 20 minutes upon its addition.

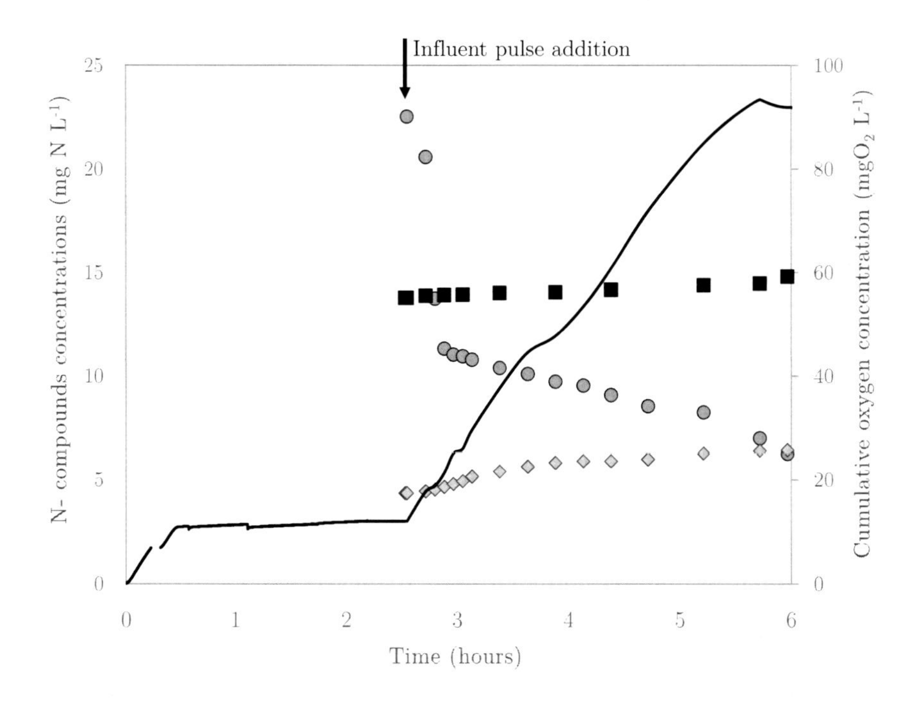

Figure 7.9. Results of the RT-4 data measured for calibration of the expanded model. NH_4^+-N (●), NO_2^--N (◇), NO_3^--N (■), and O_2 (—).

Respirometric tests 5 and 6 (RT-5 and RT-6) (Figure 7.10) were carried out by adding a higher concentration of ammonium than in previous tests. On average, the ammonium addition was 82.6 (± 3.3) mg NH_4^+-N L^{-1}. For both tests the biomass was left overnight for 12 hours without any feeding and under light conditions in order to exhaust any nitrogen stored intracellularly. In both tests, nitrification removed in average 50.0 (± 12.3) % and algal uptake 50 (± 12.7) % of the ammonium available. Also, for both tests the calculated oxygen production was 0.023 gO_2 L^{-1} h^{-1}.

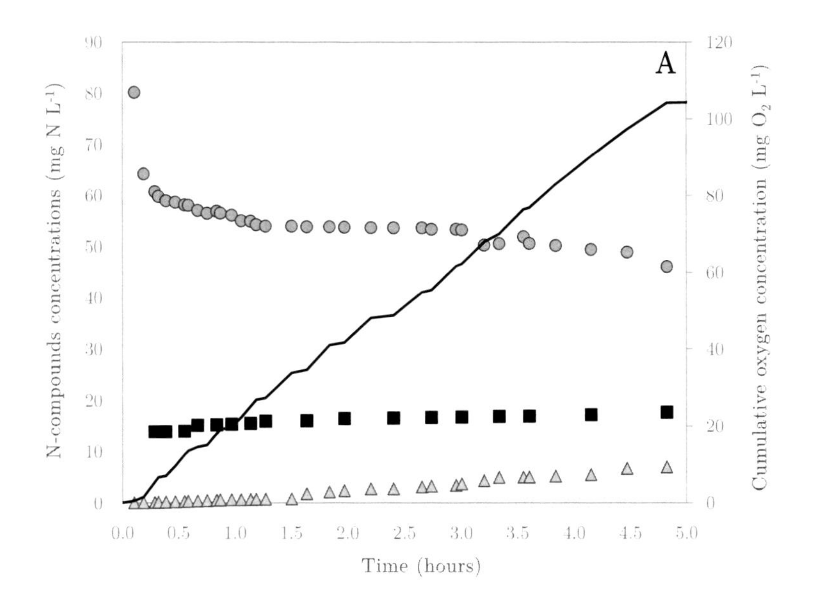

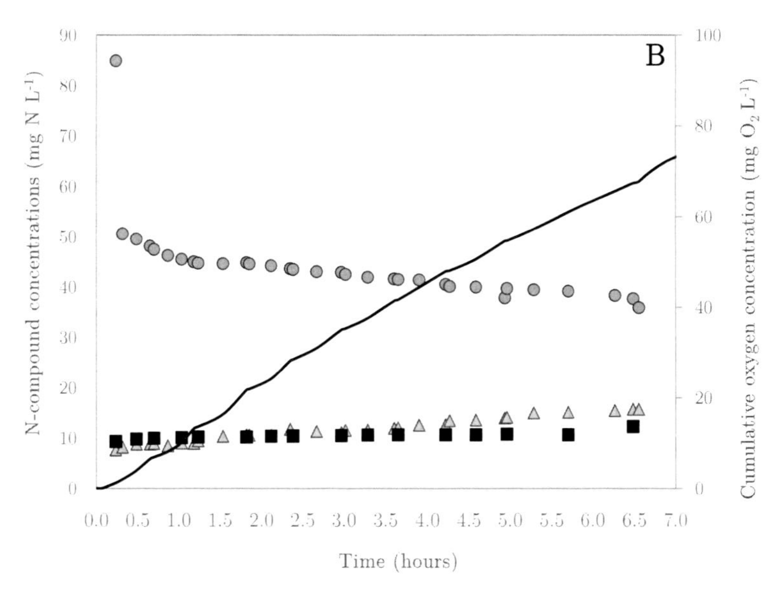

Figure 7.10. Results of RT-5 (A) and RT-6 (B) measured for calibration of the expanded model. NH_4^+*-N (●),* NO_2^-*-N (◇),* NO_3^-*-N (■), and* O_2 *(▬).*

Model calibration

The model was calibrated using the data sets obtained in three respirometric tests (RT-4, RT-5 and RT-6). The model was successfully calibrated being able to describe the following processes: nitrogen storage, phototrophic growth on intracellularly stored nitrogen, phototrophic growth on dissolved nitrogen in the bulk liquid, and ammonium oxidation by nitrifiers. The growth of heterotrophic bacteria (organic carbon oxidation and denitrification) was not calibrated, as no COD was added in the respirometric tests. The calibrated values for all three tests for the nitrogen storage and the growth of algae on N-stored were: $f_{n,max}$, $k_{sto,N}$, $K_{STO,N}$, $\mu_{STO,N,P}$, $K_{STO,N,P}$, and $f_{X_N}^{REG}$. Furthemore, for nitrification the calibrated parameters were: μ_P μ_{AOB}, and μ_{NOB}. Table 7.1 presents the results of the parameters calibrated for the processes previously listed. Figure 7.11 and Figure 7.12 present the comparison between the modelled and measured data. As seen from the calculation of the error (Table 7.2), the results of the model could describe accurately the trends of the conversions of ammonium removal, and nitrite, nitrate and oxygen production during the respirometric tests.

Table 7.1. Calibrated parameters for the nitrogen storage, phototrophic growth on both nitrogen storage and external ammonium, and autotrophic processes for RT-4, RT-5 and RT-6.

Parameter	*RT-4*	*RT-5*	*RT-6*	*Unit*
		Nitrogen storage process		
$k_{sto,N}$	20	20	20	d^{-1}
$K_{STO,N}$	0.0001	0.0001	0.0001	$g\ N\ gCOD_{X\_P}^{-1}$
$f_{n,max}$	0.13	0.22	0.35	$g\ N_{sto}\ g\ COD_{X\_P}^{-1}$
		Phototrophic growth on $\boldsymbol{X_N}$		
$f_{X_N}^{REG}$	0.009	0.005	0.09	$g\ N_{sto}\ g\ COD_{X\_P}^{-1}$
$\mu_{STO,N,P}$	1.2	1.2	1.2	d^{-1}
$K_{STO,N,P}$	0.2	0.2	0.2	$g\ N_{STO}\ g\ COD_{XP}^{-1}$
		Phototrophic growth on $\boldsymbol{S_{NH_4^+}}$ and autotrophic growth		
μ_P	3.5	3.5	3	d^{-1}
μ_{AOB}	0.22	0.5	0.5	d^{-1}
μ_{NOB}	0.32	0.76	0.71	d^{-1}

Table 7.2. IOA calculated between the modelled and measured data for the different compounds for RT-4, RT-5 and RT-6.

Compound	***RT-4***	***RT-5***	***RT-6***
	IOA value		
$S_{NH_4^+}$	0.88	0.91	0.87
$S_{NO_2^-}$	0.98	0.98	0.87
$S_{NO_3^-}$	0.97	0.80	0.94
O_2	0.98	0.99	0.99

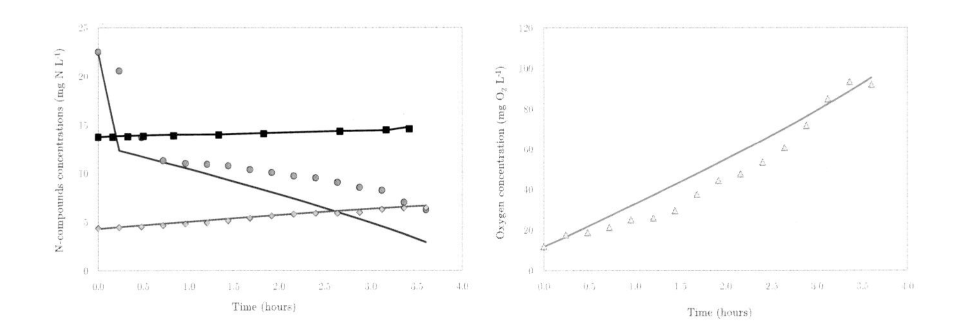

Figure 7.11. Calibration results for RT-4. Measured NH_4^+-N (●), measured NO_2^--N (◆), measured NO_3^--N (■), measured O_2 (△), modelled NH_4^+-N (▬), modelled NO_2^--N (▬), modelled NO_3^--N (▬), and modelled O_2 (▬).

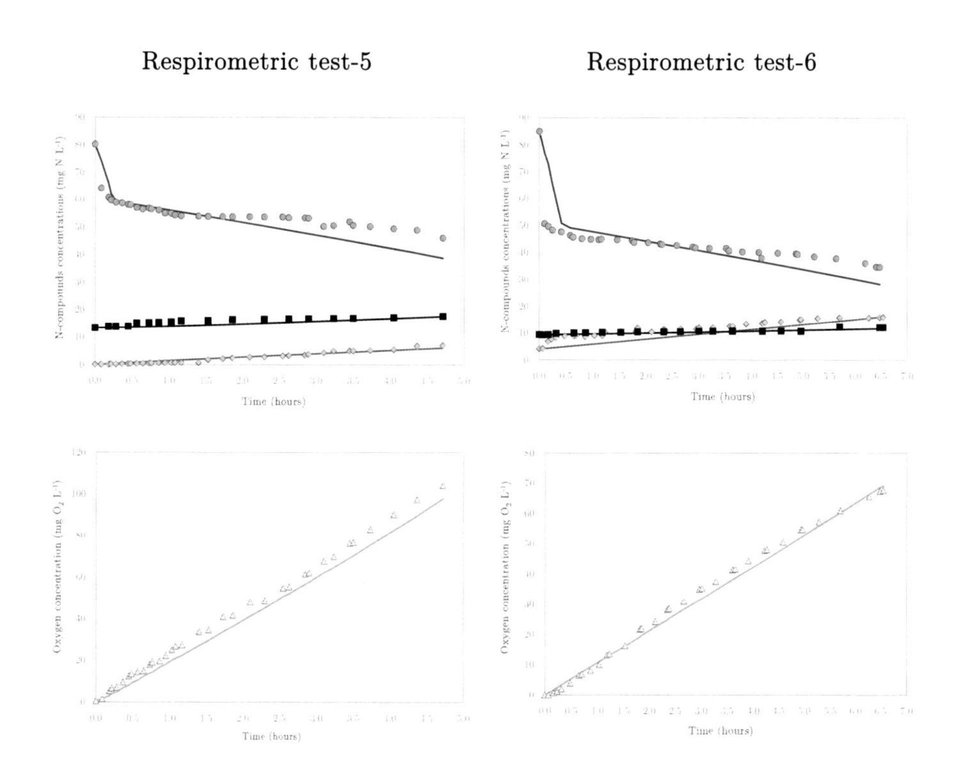

Figure 7.12. Calibration results for RT-5 and RT-6. Measured NH_4^+-N (●), measured NO_2^--N (◇), measured NO_3^--N (■), measured O_2 (△), modelled NH_4^+-N (▬), modelled NO_2^--N (▬), modelled NO_3^--N (▬), and modelled O_2 (▬).

To describe the experimental data, the maximum nitrogen storage capacity of each test was found. For RT-6, it had a value of 0.33 g N_{sto} $gVSS_{X\_P}^{-1}$, while the lowest was found for RT-4 (0.12 g N_{sto} $gVSS_{X\_P}^{-1}$). These values are similar to those reported by Flynn et al. (1993), who observed a maximum nitrogen storage within the cell of 0.2 g N per g algal biomass. However, these values are higher than the ones calculated based on the measured data. This may be related to the fact that the tests used for modelling presented a higher ammonium concentration in the pulse addition. Also, the nitrogen storage calculations performed are based on the stoichiometry assuming that all the oxygen produced proceeded from the uptake of nitrogen from the bulk liquid, implying that there was less nitrogen to be stored within the cells. In this study, it was found that the $f_{X_N}^{REG}$ is higher when the storage of ammonium is maximized. This parameter regulates the amount of intracellularly stored nitrogen inside the cell that is used for growth. However, this value presents a high variability (ranging from 0.009 to 0.09 g N_{sto} g $VSS_{X\_P}^{-1}$) due to unknown reasons. Likely, it depends on the maximum storage N capacity, but the cell growth process and factors that affect the intracellular N storage utilization may have a strong effect on this parameter. Therefore, more studies are necessary to assess the factors that regulate the utilization of the intracellularly stored nitrogen pools. The growth rate of algae on the stored nitrogen remained similar in all the three tests assessed (at 1.2 d^{-1}), while the highest growth rate on the ammonium present in the medium was 3.5 d^{-1}. Wágner et al. (2016) observed algal growth rates of between 3.54 and 4.12 d^{-1} in a culture composed mainly of *Chlorella sorokiniana* and *Scenedesmus*. The rate at which nitrogen was stored in the cell was the same for all three tests (20 g N g $COD_{X\_P}^{-1}$ d^{-1}), which is higher than the one reported by Wágner et al. (2016) of 0.36 gN $gCOD^{-1}$ d^{-1}.

The algal growth model proposed by Droop (1973, 1983) differentiates the nitrogen uptake from the nitrogen used by growth, by introducing an extra "compartment", and proposing a luxury uptake of nutrients, in this case nitrogen. The storage of nitrogen occurs in all phytoplankton species in different environments (Lavín and Lourenço, 2005). In addition, other authors have reported the storage of nitrogen by microalgae (Lavín and Lourenço, 2005; Mooij et al., 2015; Wágner et al., 2016).

The ability of algae to store nitrogen allows the growth and maintenance of the microorganisms when nitrogen has been depleted (Flynn, 1990; Wágner et al., 2016). It has been reported that the storage of nitrogen occurs in diatoms under nitrogen limiting conditions (Mooij et al., 2015), and in phytoplankton in which the nitrogen uptake will depend on the nutrient concentration in the medium, while the growth is associated to the internal concentration (algal tissue) (Fong et al., 1994). Furthermore, the internal nitrogen pool of the microalgal cell can affect the specific nitrogen uptake rate by algae, which decreases when the internal nitrogen concentration is high (Quinn et al., 2011). Nevertheless, in this study, a similar approach like that propose by Droop (1973) was followed by including the intracellular storage of nitrogen and its further utilization by algae. This approach satisfactorily described the storage processes and the nitrogen and algal-biomass activity observed in three different experiments. Since the mechanisms that trigger the intracellular nitrogen storage mechanisms are not fully clear, more studies are necessary to explain the internal pathways associated to the storage of nitrogen (Mooij et al., 2015). The model developed in this study can nevertheless be used as tool to assess in more detail the required mechanisms and contribute to get a better understanding of this process.

A deeper understanding of the factors affecting the nitrogen storage within the algae cells will allow to suggest operational techniques envisioned to maximize the nitrogen uptake, while opening a wide range of possibilities for by-product recovery. By introducing dark and light feed regimes, algae can produce storage compounds such as starch, proteins, glucose or lipids depending on the feeding times of the nutrient and inorganic carbon. However, this would mean that compounds such as inorganic carbon and ammonium are fed at different times (Mooij, 2013), which is not feasible when treating wastewater. Nevertheless, as reported in this chapter, nitrogen storage occurs in microalgal-bacteria consortia at higher rates, and contributes to increasing the ammonium removal rates. Therefore, more research is needed to further validate this model and analyse other factors, such as the effect of inorganic carbon or ammonium feeding under dark conditions.

7.4 CONCLUSIONS

Respirometric tests were used for the evaluation of the ammonium removal mechanisms by a microalgal-bacterial consortia. The RTs showed the large effect of nitrogen storage by algae, since even when no ammonium was fed, a cumulative oxygen production reached maximum values of 174 mg O_2 L^{-1}. Both the ammonium removal by nitrification and algal uptake reached 50% each. Therefore, in order to evaluate the nitrogen storage by algae and the nitrification process, the model presented in Chapter 6 was expanded for N-storage and successfully calibrated, including the processes of nitrogen storage by algae, and algal growth on nitrogen stored. The maximum storage of nitrogen calculated by the model was 0.33 g N_{sto} $gVSS^{-1}$ of algal biomass. The model, that included nitrogen storage, phototrophic growth on intracellular nitrogen and growth on ammonium in the bulk liquid, was successfully calibrated. The updated model can serve as a tool to evaluate the nitrogen storage by algae in microalgal-bacterial consortia.

8

Conclusions and recommendations

8.1 Introduction

This thesis focuses on the use of microalgal-bacterial consortia for nitrogen removal. Furthermore, the use of the microalgal-bacterial consortia in an innovative system called Photo-Activated Sludge (PAS) was evaluated. The treatment is based on the symbiosis between microalgae and aerobic bacteria, in which the objective is to maximize the oxidation of ammonium and organic carbon using the oxygen produced by microalgae through photosynthesis. One of the targeted effluents for the application of these consortia are the effluents from anaerobic digesters. These effluents exhibit high concentrations of organic carbon (COD), ammonium, and phosphorous. The concentrations of ammonium can be between 400 -1150 mg NH_4^+-N L^{-1}, while for phosphorous and COD, the range is between 29 - 74 mg P L^{-1} and 920 - 7800 mg COD L^{-1}, respectively (Dębowski et al., 2017). Although microalgal systems can be an economic and sustainable option for the treatment of these effluents, the large areas (Rawat et al., 2011) and hydraulic retention times required for their operation present an important drawback for this technology. Furthermore, the low settling characteristics of the microalgal biomass increase the operational costs due to the high energy consumption needed for the harvesting, and additionally decrease the possibility of biomass recovery for further production of by-products (Christenson and Sims, 2011). Microalgal-bacterial consortia have shown promising results in both aspects: increasing the ammonium removal rates when treating ammonium high strength wastewater (Wang et al., 2015) and increasing the settleability of algal biomass by the addition of bacteria (Quijano et al., 2017; Tiron et al., 2017; Van Den Hende et al., 2014). However, the optimal operational parameters to maximize the removal rates and/or efficiency and the factors that influence the different removal mechanisms (de Godos et al., 2016), as well as the metabolic interactions between these two groups of microorganisms are still not fully understood.

Therefore, during this research the main objective was to assess how the dual action of microalgae and aerobic bacteria could successfully treat these effluents by maximizing the ammonium removal rates and at the same time quantify the different removal mechanisms and interactions occurring in the microalgal-bacterial systems. The research was divided in four major sections: the first section (Chapter 3 and 4) assessed the removal mechanisms in microalgal-bacterial consortia using synthetic wastewater under different operational conditions in two types of photobioreactors (circular and flat panel reactor). The second section (Chapter 5) consisted of the assessment of the microalgal-bacterial consortia using real wastewater (effluent from an anaerobic digester treating swine manure). In the third section (Chapter 5 and 6), the calibration and validation of a mathematical model that describes the microbiological processes occurring between aerobic oxidising bacteria (AOB), nitrite oxidising bacteria (NOB), heterotrophic bacteria and microalgae. In the fourth section (Chapter 7), the use of respirometric tests for microalgal-bacterial biomass showed the fate of the nitrogen taken up by microalgae and its effects on the nitrification. Figure 8.1 shows the most important findings per chapter, and how the information found was used from chapter to chapter. A further discussion of each of these findings can be found in the sections below.

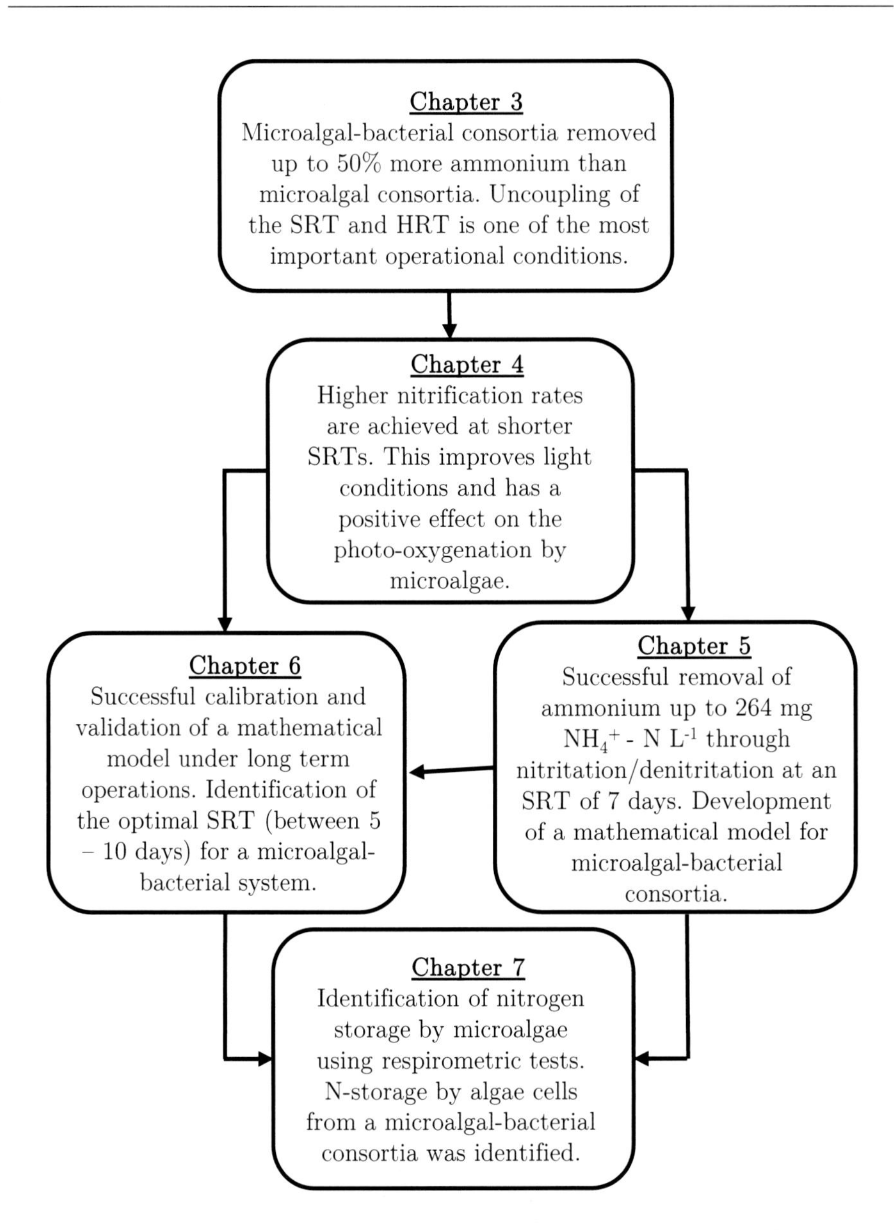

Figure 8.1. Key important findings of the research on algal-bacterial systems performed in this PhD study.

8.2 ADVANTAGES OF MICROALGAL-BACTERIAL CONSORTIA FOR AMMONIUM REMOVAL

8.2.1 Advantages on ammonium removal rates

In Chapter 2, it was demonstrated that microalgal-bacterial consortia removed ammonium 50% times faster than in a solely microalgal system, which ultimately increases the efficiency of the system. Furthermore, Chapter 3 had the highest ammonium removal rate and specific ammonium removal rate in comparison with the other chapters in this research (Table 8.1). The main removal mechanism that contributed to the increase in the ammonium removal rates was nitrification. Furthermore, other studies have also reported the successful treatment of high strength wastewater using microalgal-bacterial cultures (Godos et al., 2010; González et al., 2008; Wang et al., 2015; Zhao et al., 2014). The removal rates obtained during this research are higher than those reported by solely algal cultures treating a diverse range of ammonium concentrations in the influent (Abou-Shanab et al., 2013; Aslan and Kapdan, 2006; Cabanelas et al., 2013). Furthermore, as stated in the general objectives of this thesis (See section 2.3), the algae strains used as inoculum were a combination between eukaryotic algae and prokaryotic cyanobacteria. Yet, once the reactors reached steady state, the most predominant algal strain was *Chlorella*. In the literature, it can be found that the most used strains of microalgae for wastewater treatment are *Chlorella sp.* (Cabanelas et al., 2013; Ruiz et al., 2011), *Scenedesmus sp.* (Kim et al., 2013; Park et al., 2010) *and Spirulina sp.* (Olguín, 2003).

Table 8.1. Summary of volumetric and specific ammonium removal rates under the different operational conditions tested in each chapter

Chapter	*Influent (mg NH_4^+ L^{-1})*	r_{Am_T} *(mgNH_4^+-N L^{-1} h^{-1})*	k_{Am_T} *(mgNH_4^+-N mgVSS^{-1} d^{-1})*	*SRT (d) & HRT (d)*	*Light intensity (µmol m^{-2} s^{-1})*
3	297.3	4.16 ± 0.75	1.84 ± 0.12	SRT:4.2 ± 0.3 HRT: 1	700
4	23	2.12	0.063 ± 0.009	SRT:17 HRT: 0.5	25.9
5	264 ± 10	2.4 ± 0.17	0.033 ± 0.002	SRT:7 HRT: 4	84±3
7	45.36 ± 5.52	3.21 ± 0.24	0.063 ± 0.012	SRT:10 HRT: 1	766.5 ± 154.1

r_{Am_T}: Volumetric ammonium removal rate; k_{Am_T}: specific ammonium removal rate

The presence of nitrifiers in the microalgal culture increased the volumetric and specific ammonium removal rates. The oxidation of ammonium by nitrifiers is faster than the algal uptake (Chapter 5). Therefore, the presence of nitrifiers in the biomass has a strong impact on the removal of ammonium despite they have a low content in the total biomass composition, between 1.8 to 17% (Chapter 3, 4 and 5). Also, the presence of other microorganisms played an important role in the total nitrogen removal. For instance, heterotrophic bacteria not just removed the organic carbon present in the influent, but also removed ammonium for their biomass growth (Chapter 4 and 5). In addition, during anoxic periods, heterotrophic bacteria, when sufficient organic carbon is present, could denitrify the nitrate or nitrite produced by nitrification (Chapter 4 and 5).

8.2.2 Operational conditions and area requirements

In Chapter 3, the ammonium removal rate by the reactor containing just microalgae was 1.84 ($\pm$ 0.66) mg NH_4^+-N L^{-1} h^{-1} (See section 3.4) and the specific ammonium removal rate was 0.025 ($\pm$ 0.009) mg NH_4^+-N $mgVSS^{-1}$ d^{-1}. These values are significantly lower than the results for microalgal-bacterial reactors in the remaining chapters (Table 8.1). Thus, for 100% ammonium removal in the microalgal reactor of Chapter 3, and assuming that the volumetric ammonium removal would remain similar, the required HRT would be approximately 6.7 days, assuming all other macronutrients and micronutrients are sufficient. Alcántara et al. (2015) calculated that in a microalgae-based system, such as high rate algae ponds (HRAP) treating medium-strength domestic water, the necessary HRT would be 7.5 for complete nitrogen and phosphorous removal. Higher nitrogen uptake by algae would result in a higher concentration of solids, which limits the light penetration, and thus reduces the growth rate of algae. Noteworthy, HRT values in HRAP could be reduced when carbon dioxide is sparged to avoid inorganic carbon limitation. This can also help as a pH-control to maintain an optimum pH. Park and Craggs (2011) obtained ammonium removal efficiencies of up to 83.3% at a HRT of 4 days with CO_2 addition in a high rate algae pond treating an effluent from anaerobic digestion. However, in HARPs with CO_2 supply, the growth of nitrifiers can be enhanced, especially when inorganic carbon is not limiting and in most cases when the HRT is not long enough for nitrifiers to grow (de Godos et al., 2016; Park and Craggs, 2011). The latter occurs in conventional HRAPs where the HRT and the SRT are not uncoupled and therefore the HRT corresponds to the solids retention time (SRT).

The high ammonium removal rates (volumetric and specific) by microalgal-bacterial consortia can further help to reduce the HRT of the system. This can be done by ensuring that the main ammonium removal mechanism within the microalgal-bacterial system is through nitrification. Comparing the oxygen production by algae with the oxygen consumption by nitrification, the yield of oxygen on ammonium consumed is 16.85 gO_2 gNH_4^+-N^{-1} consumed (Mara, 2004). This is significantly higher than the 4.57 gO_2 gNH_4^+-N^{-1} required for complete nitrification (Ekama and Wentzel, 2008a). Therefore, the design of a microalgal-bacterial system should ensure enough oxygen production by algae to support all aerobic processes. Another important condition that should be met is the retention of nitrifiers within the system. Thus, for the cultivation of a microalgal-bacterial consortium in which nitrification is envisioned as the main removal mechanism, there should be an uncoupling between the SRT and the HRT (Chapter 3; Valigore et al., 2012).

The possibility of reducing further the HRT by the uncoupling between the SRT and HRT in a microalgal-bacterial system has positive effects on the nitrification process, and the objective of microalgae supplying the necessary oxygen to support the aerobic processes. Also, the reduction of the HRT contributes to the reduction of the large area requirements of algal systems. Since microalgae would not be the main removal mechanisms, the limitation of light by solids should be enough to support photo-oxygenation. Therefore, the designing depths of reactors using microalgal-bacterial consortia could be deeper. During Chapter 3, the microalgal-bacterial system had a surface removal rate of 10.2 g $NH_4^+ - N$ m^{-2} d^{-1}, compared with 4.4 g $NH_4^+ - N$ m^{-2} d^{-1} for the microalgal consortia. Comparing these values with the study of Tuantet et al. (2014), who achieved a maximum removal rate of 54.1 mg $NH_4^+ - N$ L^{-1} h^{-1} using urine as growth medium, the surface ammonium removal rate calculated was 6.5 g $NH_4^+ - N$ m^{-2} d^{-1}. This value is lower than for microalgal-bacterial systems, and also the reactor used for cultivation by Tuantet et al. (2014) had a short light path of 5 mm, which avoided any light limitation in the culture. In practice, HRAP are designed with a HRT between 2 to 8 days and depths between 0.2 to 0.5 m (Shilton, 2006). Using the information reported by Park and Craggs (2011) in a HRAP treating domestic wastewater, the surface removal rate was estimated to be 1.1 g $NH_4^+ - N$ m^{-2} d^{-1}, which is considerably lower than the values found in this thesis. In summary, the uncoupling of the HRT and SRT allows to develop a higher settleable biomass. Consequently, both SRT and HRT can be further shortened, which has a positive result on the light limitation by solids and on the nutrient removal rates. As a result, the depth (light path) of the reactors using microalgal-bacterial consortia, in which the main ammonium removal mechanism is through nitrification, can be further decreased, which will help to reduce area requirements. Based on the results presented in this thesis (see section 3.3.5), the area requirements for a microalgal-bacterial consortia can be reduced up to 50% in comparison with solely algal systems. Nonetheless, the rates presented in this research are calculated based on laboratory-scale experiments, and more research is required at pilot scale in order to define minimum depths that are able to meet the necessary oxygen production, and at the same time maintain the nutrient removal efficiency of the system.

8.2.3 Photo-oxygenation and algal harvesting

Another important advantage of the use of microalgal-bacterial consortia over other technologies are the economic costs. Especially on two aspects: the cost of aeration when comparing this technology with activated sludge, and the cost of harvesting when comparing with algal systems. Comparing this technology with activated sludge systems, the oxygen required for nitrification and COD oxidation is fully supported by microalgae (Chapter 3, 4 and 7). Operational costs by aeration can represent up to 60 to 80% (Holenda et al., 2008) of the total operational costs in activated sludge plants. The energy consumption is on average between 0.33 to 0.60 kWh m^{-3} in activated sludge plants in the United States (Plappally and Lienhard, 2012), while for HRAP the power consumption for mixing, calculated by Alcántara et al. (2015), was 0.023 kWh m^{-3}. Therefore, the energy needed for removal of ammonium in high strength wastewater using an activated sludge process would be considerably higher when compared with a microalgal-bacterial system.

Another advantage of the microalgal-bacterial systems is the improvement in the settling characteristics of the biomass (Chapter 4 and 5) when compared with algal systems. The uncoupling of the SRT and HRT, and the operation in sequencing batch creates a selective environment for fast settleable microalgae, and furthermore promoted the formation of algal-bacterial aggregates. This positive effect on biomass harvesting by the presence of bacteria in algal systems has been reported by other studies as well (Gutzeit et al., 2005; Park and Craggs, 2011; Van Den Hende, 2014). Furthermore, the increase in settleability reduces the cost of operation in these systems, and so no extra energy is required for solids separation, such as centrifugation or dissolved air flotation. In addition, the bioflocculation avoids contamination of the biomass, since no chemicals are needed to promote flocculation (Su et al., 2011). Finally, more studies are under development to improve this positive effect of algae and bacteria. For instance Tiron et al. (2017) published an approach to develop activated algae granules which have sedimentation velocities of 21.6 (± 0.9) m h^{-1}, and in terms of the separation of the algal biomass from the bulk liquid, the biomass recoveries were up to 99%.

8.3 Influence of the SRT on the operation of a microalgal-bacterial photobioreactor

Chapter 3 showed that the uncoupling of the SRT and HRT is imperative for the development of a steady nitrifying microalgal-bacterial consortium. Furthermore, Chapter 4 and 5 showed the effects of the SRT on the removal mechanism of microalgal-bacterial consortia, still the ammonium removal efficiency was 100% under the different operational conditions tested. In both chapters, volumetric and specific ammonium removal rates were higher at shorter SRTs (17 days SRT for Chapter 4 and 7 days SRT for Chapter 5). Furthermore, the ammonium removal mechanisms differ at different durations of the SRT. In Chapter 4, at a longer SRT of 52 days, ammonium removal by algal uptake represented up to 38% of the total ammonium removal, while it decreased up to 11% at a SRT of 17 days (Table 4.3). In both cases, the main ammonium removal mechanism was nitrification/denitrification.

Therefore, one of the most important operational parameters to control the efficiency and rates of ammonium removal in microalgal-bacterial consortia is the SRT. The SRT controls the amount of solids in the reactor, which will have a high impact on the light penetration used for algal growth and consequently oxygen production. Longer SRTs in activated sludge increase the concentration of endogenous residues, which reduce the active fraction of the biomass and increase the oxygen consumption through respiration of the bacterial biomass (Ekama and Wentzel, 2008b). In addition, longer SRTs increase the solids concentration in the reactor, hence the dark zones within the reactor increase (Figure 5.), which will also increase the oxygen consumption by algal respiration. As a result, oxygen is less available for the aerobic processes such as organic carbon oxidation and nitrification, resulting in a shift in the removal mechanism from nitrification to algal uptake. However, if the HRT is not long enough and the ammonium concentration in the influent is high, the efficiency of the system could be hindered, and both high concentrations of nitrite and ammonium (partial nitrification and no denitritation) and organic carbon can end up in the effluent.

The uncoupling of the SRT from the HRT permits to select an optimum SRT that allows enough light penetration to maximize the nitrification rates and reduce the solids concentration. This will decrease the endogenous residue by the bacterial biomass, while at the same time increase the growth rate of the nitrifiers (Ekama and Wentzel, 2008a). Decreasing of the SRTs and increasing the ammonium removal rates can help to further decrease the HRT, which would as well offer the possibility to reduce the area requirement of the technology as stated above. However, HRTs shorter than 0.5 days were not tested in this research. Therefore, more research is needed to demonstrate the feasibility of this low HRT. Furthermore, it is imperative to not fall below the SRT_{min} for nitrifiers, since below this value nitrifiers would be washed out of the system and the system would collapse. Finally, based on the experiments performed during this research, the optimum SRTs for microalgal-bacterial reactors would be between 5 - 10 days (Chapter 4, 5 and 6).

8.4 Evaluation of the microalgal-bacterial consortia using mathematical models

Chapters 5, 6 and 7 present the successful calibration and validation of a mathematical model for microalgal-bacterial systems. The model describes the different microbiological processes occurring within the consortia when treating ammonium rich wastewaters. The model was proposed in Chapter 5, and is based on the modified activated sludge model number 3 (modified ASM-3) proposed by Iacopozzi et al. (2007) and Kaelin et al. (2009). The phototrophic growth on light limitation was based on a similar approach proposed by Martinez Sancho et al. (1991). In Chapter 5, the model was calibrated using the data from the hourly cycles (Figure 5.) under batch conditions. In Chapter 6, the model was calibrated and validated for long-term sequencing batch operation using the information of the hourly cycles measured during the three operational periods in Chapter 4. Finally, in Chapter 7, the model was updated and calibrated, adding two new processes to the model: nitrogen storage by microalgae and growth on this stored nitrogen (See section 7.2.3).

8.4.1 Mathematical model for analysis of phototrophic growth, nitrification/denitrification and organic carbon removal processes

The mathematical model proposed in this research could describe the measured data, reporting good values of the index of agreement (0.5 - 1) (Table 7.2). The index of agreement compared the variances between the measured data and the modelled values. The light extinction coefficient was found to be one of the most important parameters in the microalgal-bacterial system, and the more sensitive parameter during the calibration process (Chapter 6). In the BIO-ALGAE model proposed by Solimeno et al. (2017), the light factor (which includes photoinhibition, photolimitation and light attenuation) was the main limiting factor for algal growth. These results are in accordance with the conclusions made on the effects of the SRT on microalgal-bacterial systems, related with the solids concentrations in the reactor. For instance, the light extinction coefficient was measured in Chapter 5 (0.0748 m^2 $gTSS^{-1}$), while the light extinction coefficient calibrated during Chapter 6 was 0.019 m^2 $gTSS^{-1}$. The higher extinction coefficient measured in Chapter 5 is presumably influenced by the turbidity of the real wastewater used in the experiment (Table 5.1), while for Chapter 6, the experiments were done using synthetic wastewater. Furthermore, the effect of the different values of the light extinction coefficient can be seen on the predicted maximum algal growth: the maximum algal growth rate was lower in Chapter 5 (Table C.2; 0.85 d^{-1}) than in Chapter 6 (Table 6.2; 2.00 $\pm$ 0.05 d^{-1}). Therefore, it is advised to measure the light extinction coefficient of the microalgal-bacterial biomass, since it would improve the veracity of the model. By measuring this parameter experimentally, the calibration is not necessary and the model would describe the different processes occurring in the reactor.

Analysing the ammonium half saturation coefficient for algal growth in Chapter 5 and 6, and comparing with the literature on algal and microalgal-bacterial models (Table 6.2), the values calibrated in both chapters are lower than the ones found in the literature. As can be seen from Figure 4.4 and Figure 5., oxygen was produced even after ammonium was completely removed or at very low concentrations. Two hypothesis could explain this: the first one would consider the uptake of ammonium at microalgal-bacterial floc level, therefore the ammonium is not measured in the bulk liquid. The second one, and more likely, would be the storage of nitrogen within the algae cell for later use as nitrogen substrate for growth when ammonium concentrations are zero or limiting. The storage of nitrogen in the microalgal biomass was also suggested in Chapter 3 (See section 3.3.4) and 4 (See section 4.3.4), when comparing the ammonium removed against the oxygen production in the bulk liquid. Also, during Chapter 7, the performance of the respirometric tests showed the ability of algae to produce oxygen with little to no ammonium present in the bulk liquid (Figure 7.8). Furthermore, nitrogen storage by microalgae occurs naturally in phytoplankton in natural environments (Lavín and Lourenço, 2005).

The maximum growth rate of AOB and NOB, calibrated and validated in Chapter 6, was similar to values of the maximum growth rate reported in the literature (Table 6.2). Therefore, the nitrifiers did not show any sign of inhibition by the presence of microalgal biomass. The same conclusion was reported in Chapter 3 (Figure 3.5) based on the comparison between the microalgal-bacterial reactor, and the microalgal reactor. However, the inhibition of the maximum growth rates of algae and bacteria in microalgal-bacterial systems treating effluents rich in ammonium could be more associated with the inorganic carbon limitation or ammonia inhibition at higher ammonium concentrations and slight changes in the pH values.

Overall the model provided insight in the different interactions between microalgae and bacteria. The calibration and validation of the model in sequencing batch operation, which took into account the hydraulics and sludge wasting, served as a tool for the evaluation of further scenarios that were not tested experimentally. The model found that the optimum SRT lies between 5 to 10 days. This tool can be further improved to include more processes such as growth of algae on nitrate or transfer of inorganic carbon and oxygen from the atmosphere into the photobioreactor.

8.4.2 Respirometric tests and mathematical model for the analysis of nitrogen storage by microalgae

The results of the respirometric tests reported in Chapter 7 showed a high and rapid uptake of ammonium by algae. In addition, the relation between the ammonium taken up by microalgae and the oxygen produced were not in balance. It was demonstrated that one of the factors that would force algae to store nitrogen was related to the limitation by inorganic carbon (Figure 7.7). Therefore, algae would store the ammonium in intracellular pools for later use when inorganic carbon is present and ammonium is limited. It has been reported that under nitrogen limiting conditions, phytoplankton stores nitrogen (ammonium, nitrate and rare occasions nitrite) within the cell, and furthermore in cultures under nitrogen-starved conditions, nitrogen uptake can be faster than when it is assimilated for growth (Dortch et al., 1984). This rapid uptake is stored in the so called transient pools (Dortch et al., 1984). However, more research is needed to understand the internal metabolic processes (Mooij et al., 2015) and how the photosynthetic apparatus changes when algae store nitrogen within the cell.

Based on the results of the respirometric tests, the model was updated by adding two processes: nitrogen storage by algal biomass, and phototrophic growth on the stored nitrogen. The introduction of the two processes resulted in the identification of two important parameters: the maximum amount of nitrogen stored per gram of algal biomass, and the regulation factor within the cell for the use of stored nitrogen for phototrophic growth ($f_{X_N}^{REG}$). The maximum nitrogen storage capacity was 0.33 g N_{sto} gVSS^{-1}. The maximum storage capacity of the cells is influenced by the algal strain, the nitrogen compound fed, and whether the culture is under conditions of nitrogen limitations, or the opposite (Dortch et al., 1984). During these experiments, it was also found that inorganic carbon could also trigger the nitrogen storage. During the proposition of the model, it was assumed that the utilization of the stored nitrogen would occur at a lower rate, and this could occur simultaneously, which is highly dependent on the regulation factor within the cell. The calibration of the respirometric tests resulted in different values of the internal regulation factor ($f_{X_N}^{REG}$) among the different calibrated respirometric tests (Table 7.1), hence the high variations made it risky to conclude about this parameter. Instead, more research is required in order to fully understand the factors that govern nitrogen storage in microalgal-bacterial consortia.

8.5 Outlook and concluding remarks

The findings in this PhD dissertation show that microalgal-bacterial consortia are able to effectively remove nitrogen at shorter SRTs and HRTs than usually used in algal systems, showing high ammonium removal efficiencies. Furthermore, the co-cultivation of microalgae and bacteria offers advantages such as higher ammonium removal rates through nitrification/denitrification and consequently reduction of the area requirements in the implementation of the technology. Also the development of a bioflocculant algal-bacterial biomass without the addition of chemicals nor energy input is an advantage.

The experiments performed and the conclusions proposed in this research were based on laboratory scale reactor set-ups. Therefore, the operational considerations made should be tested at pilot scale for further validation. The PAS system could fit within a holistic approach for wastewater treatment consisting of an anaerobic digester coupled with a microalgal-bacterial photobioreactor (Figure 8.2). The anaerobic digester is used for bioenergy production through a combined heat and power (CHP) system, and the high nutrient strength centrate is further treated in a microalgal-bacterial photobioreactor. The biomass produced in the photobioreactor can be returned to the anaerobic digester to increase biogas production by co-digestion with the main waste(water) streams (Wang and Park, 2015). Part of the stabilized solids from the anaerobic digester and the microalgal-bacterial reactor could be used as biosolids for fertilizer replacement, promoting a circular economy within the treatment of wastewater.

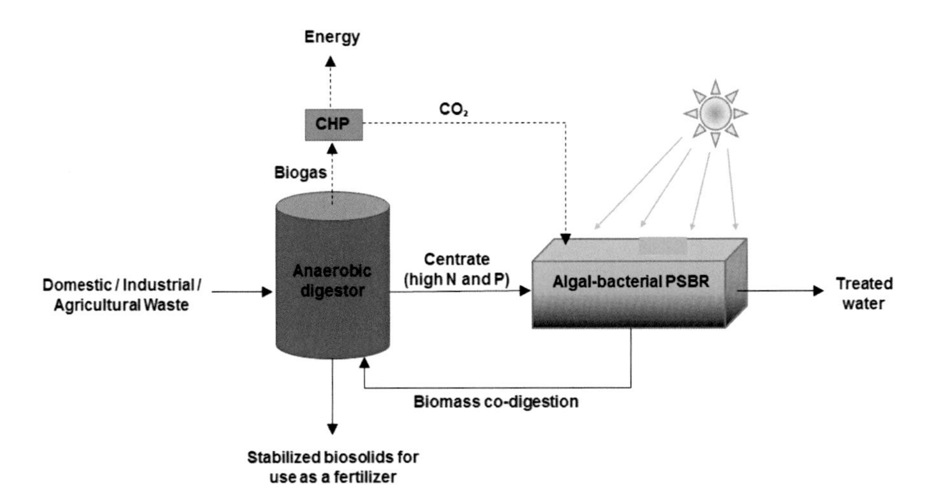

Figure 8.2. Scheme of the proposed holistic approach for treatment of domestic, industrial and agricultural wastes. CHP: combined heat and power system, N: nitrogen and P: phosphorous

At pilot scale and using sunlight as energy source, it is important to take into account the feeding conditions of the medium. However, this also depends on the final objective of the water reclamation of the treated effluent. For instance, effluents with high concentrations of nitrate, when just nitrification is performed in the microalgal-bacterial system, can support irrigation for crop growth (Taylor et al., 2018). In case that due to the prior treatment there is a lack of micronutrients or other nutrients such as phosphorous, the effluent can be mixed in a certain ratio with the influent from the anaerobic digester to supply all the compounds needed. When the objective of the microalgal-bacterial system is the treatment of the wastewater to negligible ammonium and total nitrogen concentrations, the system should support nitrification and denitrification as seen in Chapter 4 and 5. Then, during a HRT of 1 day, nitrification can be performed during the daylight and denitrification can be supported at night when there is no longer oxygen production. Therefore, it is recommended that the influent is fed during the dark conditions, then some of the oxygen still present from the light phase would be consumed for organic matter oxidation and part of the ammonium would be oxidized or taken up by algae. The rest of the organic matter would be used for denitrification, and the remaining ammonium that is not nitrified or taken up in the dark phase would be nitrified in the next light phase.

Taking into account the results from Chapter 7 on nitrogen storage by algae, the night feeding could also promote the nitrogen storage within the cells as observed by Mooij et al. (2015). Furthermore, this nitrogen feeding regime could also limit the nitrogen during the light feeding as partly will be consumed at night, which will force the algal biomass to store inorganic carbon as mainly carbohydrates and lipids (Mooij et al., 2015). Therefore, evaluation of this biomass for biofuel production could be an option. Taking into account that the growth medium is wastewater, the energy balance for the production of algal fuels could shift to an energy positive balance and make it cost competitive against fossil fuels (Sivakumar et al., 2012). However, there is still the challenge of maximizing the lipids production within the algal cells, specially when cultivated in municipal wastewater (Tan et al., 2018).

The symbiosis of microalgae and bacteria has shown promising results not just for nutrient and organic carbon removal, but for the elimination of other pollutants and contaminants from different industries as well (Rawat et al., 2011). The results and conclusions of this thesis offer new directions for research on microalgal-bacterial consortia. New studies on the co-culturing of different microorganisms for treatment of wastewater is already on-going (Mukarunyana et al., 2018; Manser et al., 2016). This shows the ability of algae to be resilient and adapt to different microbial populations and environments, and can help to further develop microalgal-bacterial consortia as sustainable approach to today's and tomorrow's wastewater problems.

A

APPENDIX A

A.1 NITROGEN MASS BALANCE

The expression proposed by Liu and Wang (2012) was used to estimate the amount of ammonia partially nitrified to nitrite (nitritation) and full nitrification:

$$NH_4^+ + 0.0225NH_4^+ + 1.38750O_2 + 0.0900CO_2 + 0.0225HCO_3^- \xrightarrow{yield} 0.0225C_5\,H_7NO_2 + NO_2^- + 104H^+ + 0.9775H_2 \qquad \text{Eq. (A.1)}$$

Nitrification (ammonium oxidized to nitrate by nitrifiers) was calculated with the following equation (Liu and Wang 2012):

$$NH_4^+ + 0.0298NH_4^+ + 1.851O_2 + 0.1192CO_2 + 0.0298HCO_3^- \xrightarrow{yield} 0.0298C_5\,H_7NO_2 + NO_3^- + 0.9702H_2O + 2H^+ \qquad \text{Eq. (A.2)}$$

Algae growth was described based on the photosynthetic activity using the equation defined by Mara (2004) (Eq. A.3).

$$106CO_2 + 236H_2O + 16NH_4^+ + HPO_4^{2-} \xrightarrow{light} C_{106}H_{181}O_{45}N_{16}P + 118O_2 + 171H_2O + 14H^+ \qquad \text{Eq. (A.3)}$$

The total removal of ammonium by nitrifiers was calculated using Eq. (A.4), and the total uptake (removal) of ammonium by algae was determined by Eq. (A.5) as follows:

$$Total\ ammonium\ removed_{nitrifiers} = S_{NO_2^- -N,effluent} * Q * \frac{1}{Y_{nitrite}} + S_{NO_3^- -N,effluent} * Q * 1/Y_{nitrate} \qquad \text{Eq. (A.4)}$$

$$Total\ ammonium\ removed_{algae} = (S_{NH_4^+-N,influent} * Q) - (S_{NH_4^+-N,effluent} * Q) - Total\ ammonium\ removed_{nitrifiers} \quad \text{Eq. (A.5)}$$

Where:

$Total\ ammonium\ removed_{nitrifiers}$: Total amount of ammonium removed by nitrifiers which mainly comprises the ammonium oxidized (in mg d^{-1}).

$S_{NO_2^--N,effluent}$: Concentration of nitrite measured in the effluent (in mg L^{-1}).

Q: Daily flow fed to the reactors (in L d^{-1}).

$Y_{nitrite}$: Ratio of nitrite produced through nitritation, calculated based on Eq (A.1) (0.977 mg $NO_2^- - N/$ mg $NH_4^+ - N$).

$S_{NO_3^--N,effluent}$: Nitrate concentration in the effluent (in mg L^{-1}).

$Y_{nitrate}$: Ratio of nitrate produced from ammonium oxidation, calculated using Eq (A.2) (0.971 mg $NO_3^- - N/$ mg $NH_4^+ - N$).

$Total\ ammonium\ removed_{algae}$: Ammonium uptake by algae for growth (in mg d^{-1}).

$S_{NH_4^+-N,influent}$: Ammonium concentration in the influent (in mg L^{-1}).

$S_{NH_4^+-N,effluent}$: Ammonium concentration in the effluent (in mg L^{-1}).

The biomass production of nitrifiers and algae in the FPRs for each cycle was calculated using their nitrogen growth requirements, based on the amount of ammonium oxidised and removed by nitrifiers and algae, respectively, using the following expressions:

$$Production_{VSS,algae} = Total\ ammonium\ removed\ by\ algae * Y_{algae} \quad \text{Eq. (A.6)}$$

And

$$Production_{VSS,nitrifiers} = Total\ ammonium\ removed\ by\ nitrifiers * Y_A \quad \text{Eq. (A.7)}$$

Where:

$Production_{VSS,algae}$: Production of algae VSS in the reactor (in mgVSS d^{-1})

Y_{algae}: Algae yield coefficient per ammonium taken of 10.83 mgVSS mg$NH_4^+ - N^{-1}$ based on Eq. (A.3).

$Production_{VSS,nitrifiers}$: Concentration of nitrifiers VSS in the reactor (in mgVSS d^{-1})

Y_A: Nitrifiers yield coefficient of 0.24 mgVSS mg$NH_4^+ - N^{-1}$ based on Eq. (A.2) (in mgVSS mg$NH_4^+ - N^{-1}$)

The nitrogen uptake by algae calculated from the nitrogen balance (Eq. A.5) was compared with the theoretical nitrogen uptake by algae. The last one was calculated using the nitrogen biomass growth requirement equation as developed by Ekama and Wentzel (2008a). The average nitrogen content within the algae biomass was calculated using the following expression:

$$N_{s,algae} = f_n \frac{V_r\, X_{VSS,algae}}{SRT} \quad \text{Eq. (A.8)}$$

Where:

$N_{s,algae}$: Theoretical nitrogen uptake by algae (in mg$NH_4^+ - N$ d^{-1})

f_n: Stoichiometric fraction of nitrogen in algae biomass (%) estimated in 0.092 mgN mgVSS$_{algae}^{-1}$ as if Eq. A.1.

$X_{VSS,algae}$: Algae biomass (in mgVSS$_{algae}$ L^{-1}) calculated using Eq. A.6.

SRT: Sludge retention time in each FPR (in d).

B

Appendix B

B.1. NITROGEN MASS BALANCE

The nitrogen balance was made in order to identify the fractions of biomass oxidised or taken up by the biomass; the fractions removed were calculated per day. The nitrogen mass balance was calculated using the detailed 24 hour cycles of each operational period. There were 4 cycles for period 1, 3 cycles for period 2A, 5 cycles for period 2B and 4 cycles for period 2C. For each cycle

B.1.1 Calculation of the total removed ammonium

The ammonium removed was calculated based on the measured ammonium concentrations in the influent and effluent:

$$S_{NH4\_T} = (S_{NH4\_INF} - S_{NH4\_EFF})\,Q \qquad \text{Eq.B.1.1}$$

Where:

S_{NH4_T}: Total ammonium concentration removed (mg NH_4^+-N d^{-1})

S_{NH4_INF}: Total ammonium concentration in the influent (mg NH_4^+-N L^{-1})

S_{NH4_EFF}: Total ammonium concentration in the effluent (mg NH_4^+-N L^{-1})

Q: Daily flow fed to the reactors (in L d^{-1}).

B.1.2 Calculation of the ammonium removed by algae, nitrifiers and OHO bacteria.

The amount of NH_4^+-N oxidized by nitrifiers ($S_{NH4\_AOB,NOB}$) was determined using the total measured nitrate formed at the end of each cycle, and the equations Eq. B.1.2 (proposed by Liu and Wang, 2012) and Eq. B.1.3. The total amount of ammonium removed by nitrifiers also includes the nitrogen required for biomass growth.

$$NH_4^+ + 0.0298NH_4^+ + 1.851O_2 + 0.1192CO_2 + 0.0298HCO_3^- \xrightarrow{yield} 0.0298C_5H_7NO_2 + NO_3^- + 0.9702H_2O + 2H^+ \quad \text{Eq. B.1.2}$$

$$S_{NH4\_AOB,NOB} = S_{NO_3^- -N\_EFF} * Q * Y_{nitrate} \quad \text{Eq. B.1.3}$$

Where:

$S_{NH4\_AOB,NOB}$: Total amount of ammonium removed by nitrifiers which mainly comprises the ammonium oxidized (in $NH_4^+ - N$ d^{-1}).

$S_{NO_3^- -N,effluent}$: Nitrate concentration in the effluent (in mg $NO_3^- - N$ L^{-1}).

$Y_{nitrate}$: Ratio of nitrate produced from ammonium oxidation, calculated using Eq. A.1.2 (1.0298 mg $NH_4^+ - N$/mg $NO_3^- - N$).

The nitrogen requirement by the heterotrophic bacteria is calculated based on the COD removed by heterotrophic bacteria. The COD concentration was measured in the influent and effluent. The total COD removed is equal to the differences between fed and measured concentrations in the effluent. The equation used (Eq. B.1.4) is proposed by Ekama and Wentzel (2008):

$$N_S = S_{COD}\, f_N\, Q\left[\frac{1-Y_{Hv}}{1+b_H SRT}(1 + f_H b_H SRT)\right] \quad \text{Eq. B.1.4}$$

Where:

N_S: Concentration of N that is incorporated into the sludge mass (mg $NH_4^+ - N$ d^{-1}).

S_{COD}: COD concentration removed (mg COD L^{-1}), oxidized and taken up by denitrification.

f_N: N content in the sludge (0.1 mg N mg VSS^{-1}).

Y_{Hv}: Specific yield coefficient for OHO (0.45 mg VSS mg COD^{-1}).

b_H: Endogenous respiration rate (0.24 d^{-1}).

f_H: Endogenous residues fraction (0.2).

Algae growth was calculated as the difference between the total ammonium removed in the system and the ammonium removed by nitrifiers and OHO growth (Eq. B.1.5):

$$S_{NH4\_algae} = S_{NH4\_T} - S_{NH4,AOB,NOB} - N_S \qquad \text{Eq. B.1.5}$$

Where

S_{NH4_algae}: Ammonium nitrogen removed by algae (mg d-1).

B.2. BIOMASS CHARACTERIZATION

The calculation of the biomass characterization allowed to distinguish the different microorganism (nitrifiers, OHO and algae) fractions within the microalgal-bacterial consortia. The nitrogen mass balance and the total organic carbon removed was used as base for the determination of the biomass characterization.

B.2.1 Biomass production for nitrifiers

Based on the total ammonium removed by nitrifiers and the nitrogen growth requirements (calculated from Eq. B.1.2), the biomass production was determined using Eq. B.2.1.:

$$MX_{VSS\_AOB,NOB} = S_{NH4\_AOB,NOB} * Y_{VSS,AOB,NOB} \qquad \text{Eq. B.2.1}$$

Where:

$MX_{VSS\_AOB,NOB}$: Biomass production per day of nitrifiers VSS (mg VSS d^{-1}).

$Y_{VSS,AOB,NOB}$: Nitrifiers yield coefficient of 0.24 mgVSS mg$NH_4^+ - N^{-1}$ based on Eq. (B.1.2).

B.2.2 Heterotrophic biomass production

Heterotrophic biomass includes aerobic heterotrophic bacteria and denitrifiers (anoxic heterotrophic bacteria). Eq. B.2.2 is taken from Ekama and Wentzel (2008b) and the calculation is based in the biodegradable organic load and the SRT of the system.

$$MX_{VSS\_OHO} = \left(Q\,S_{COD} \frac{Y_{Hv}SRT}{(1+b_H SRT)} + Q\,S_{COD} \frac{Y_{Hv}SRT}{(1+b_H SRT)}\, f_H b_H SRT \right) \Big/ SRT \qquad \text{Eq. B.2.2}$$

Where:

MX_{VSS_OHO}: Biomass production per day of OHO bacteria (mg VSS d^{-1}).

S_{COD}: The total biodegradable organics, the value taken for the calculation was the total organics removed in the reactor (COD oxidation and COD for denitrification) (mg COD L^{-1}).

A.2.3 Algae biomass production

The biomass production of algae for each cycle was calculated using their nitrogen growth requirements, which was calculated using the equation proposed by Mara (2004) and the amount of ammonium taken up by algae:

$$106CO_2 + 236H_2O + 16NH_4^+ + HPO_4^{2-} \xrightarrow{light} C_{106}H_{181}O_{45}N_{16}P + 118O_2 + 171H_2O + 14H^+ \qquad \text{Eq.B.2.3}$$

$$MX_{VSS\_algae} = S_{NH4\_algae} * Y_{VSS\_algae} \qquad \text{Eq.B.2.4}$$

Where:

MX_{VSS_algae}: Production of algae VSS in the reactor (mgVSS d^{-1}).

Y_{VSS_algae}: Algae yield coefficient per ammonium taken of 10.83 mgVSS mg$NH_4^+ - N$-1 based on Eq. (B.2.3).

A.2.3 Total biomass production

The biomass characterization was obtained based on the sum of the total biomass production per day.

$$MX_{VSS} = MX_{VSS\_AOB,NOB} + MX_{VSS\_OHO} + MX_{VSS\_algae} \quad \text{Eq. B.2.5}$$

Where:

MX_{VSS}: Total biomass production in the reactor (mgVSS d^{-1}).

The biomass distribution respecting to the total VSS within the photobioreactor was calculated as follows:

$$\%nitrifiers = {MX_{VSS\_AOB,NOB}}/{MX_{VSS}} \quad \text{Eq. B.2.6}$$

$$\%OHO = {MX_{VSS\_OHO}}/{MX_{VSS}} \quad \text{Eq. B.2.7}$$

$$\%algae = {MX_{VSS\_algae}}/{MX_{VSS}} \quad \text{Eq. B.2.8}$$

B.3. AMMONIUM REMOVAL MODELLING

In order to determine the volumetric and specific ammonium removal rate, the ammonium removal measured for the cycles in each of the period was modelled in Aquasim. The model used as a base for the algal-bacterial system was published by Arashiro et al. (2016). The equations for the nitrifiers, OHO and algae processes were used as stated by the author. In the same way, the yields and kinetics were used as defined by Arashiro et al. (2016). The only parameters that were modified were the biomass characterization, light intensity, ammonium, nitrate, nitrite and oxygen concentrations and reactor characteristics (dimensions).

A.3.1 Ammonium fitting

The fitting of the ammonium profile for period 1 (4 cycles) was done using all the cycles and taking into account that there was just one feeding per cycle for this period. The operational scheme of these periods 2A, 2B and 2C had two feeding times, therefore, the fitting of the ammonium was done for both feedings. Figure B.3.1 presented an example of a cycle during period 2B. The cycles during periods 2A, 2B and 2C had the same trend in relation with the N-compounds and oxygen concentrations. During the fitting, the parameters that were modified were the growth rate of AOB, NOB and algae and the half saturation constants of AOB for ammonium and oxygen.

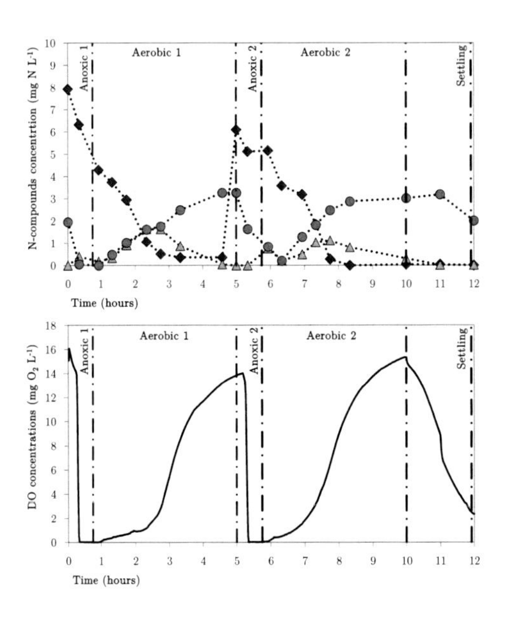

Figure B.3.1. Cycle on a day 117 during period 2B. NH_4^+-N (◆), NO_3^--N (●), and NO_2^--N (▲) and Oxygen (solid line)

During the fitting of each period, the biomass, nitrogen compound and oxygen concentrations were changed. The biomass of each microorganisms was determined using Equations B.2.6, B.2.7 and B.2.8, and the average biomass concentration for that period. For instance during period 2A, the biomass was characterized as 2.2% nitrifiers, 23.8% OHO and 74.0% algae. The average VSS concentration for that period was 2640 mg VSS L^{-1}, thus nitrifiers, OHO and algae biomass were 58.2, 627.4 and 1954.4 mg VSS L^{-1}, respectively. These values were introduced in the model in units of mg COD L^{-1}. Thus, in the case of bacteria (OHO and nitrifiers) the conversion factor used was 1.48 mg COD mg VSS^{-1} (Ekama and Wentzel, 2008b) and for algal biomass the factor was 0.953 mg COD mg VSS^{-1} (Zambrano et al., 2016). Figure B.3.2 present the fitting for the ammonium removal for the different periods.

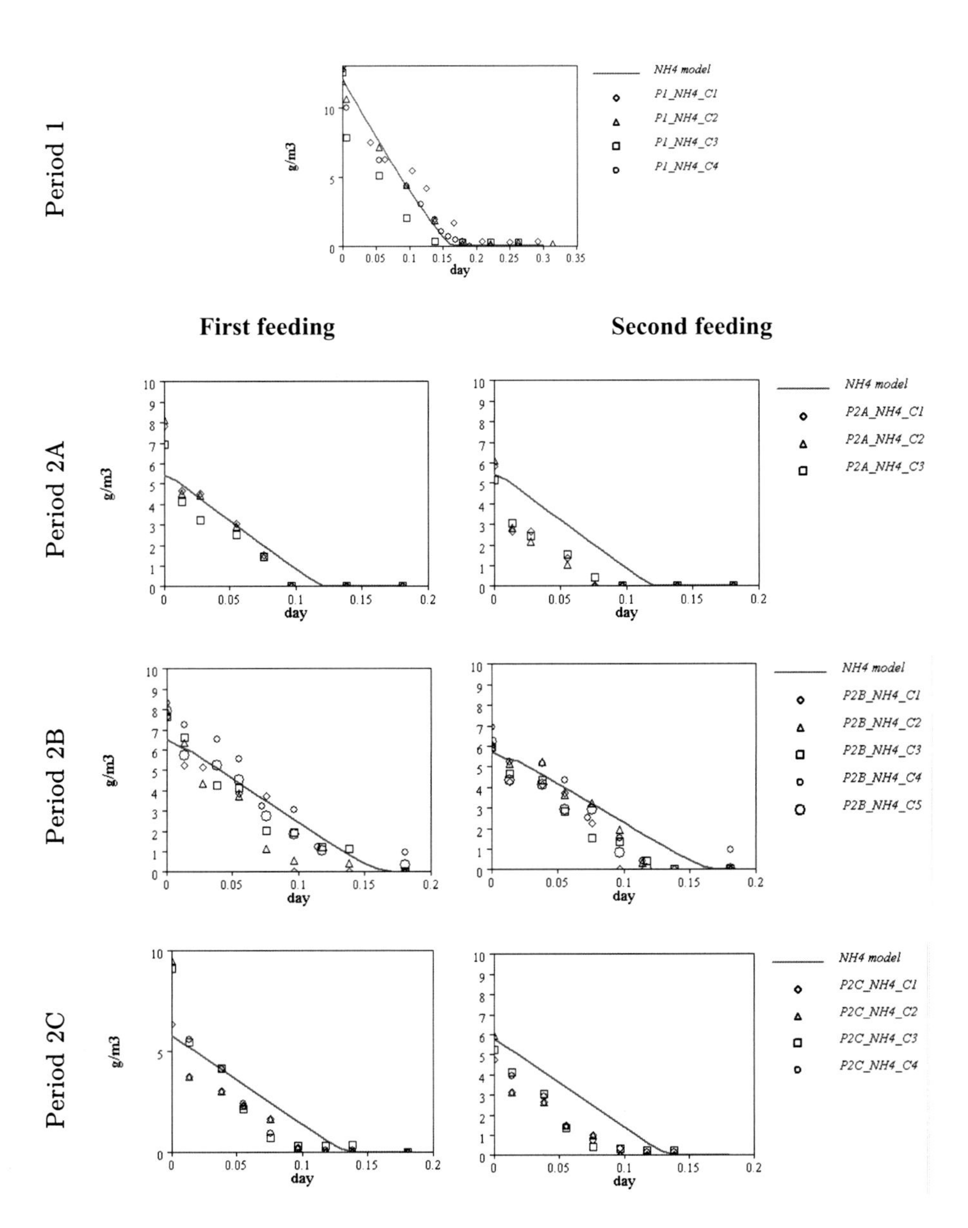

Figure B.3.2 Measured (markers) and modelled (solid line) ammonium concentration data for the different periods.

B.4. OXYGEN MASS BALANCE

The oxygen mass balance was calculated in periods 2A, 2B and 2C, since during these periods the oxygen concentration data was detailed and complete. This allows to determine the change of oxygen in a fixed period of time. Furthermore, the oxygen balance was aimed to determine the mass of oxygen production by algae and the oxygen consumption by the different aerobic processes in a day (mg O_2 d^{-1}). Therefore, the balance was calculated for the light phases of the cycles. The approach followed was:

B.4.1 Determination of the oxygen transfer coefficient of the reactor.

The oxygen transfer was measured using the same approach by Zalivina (2014) and the transfer coefficient determined was 0.48 h^{-1}.

B.4.2 Oxygen mass balance equation for determination of the O_2 produced by algae.

During the light period, considered the aerobic periods, the oxygen mass balance was defined as follows:

$$\frac{dSO_2}{dt} = KL_a(SO_{2\_S} - SO_2) + O_{2\_algae} - O_{2\_nitr} - O_{2\_OHO} - O_{2\_resp_{algae}} - O_{2\_resp_{OHO}}$$

Eq. B.4.1

Where:

SO_2: Oxygen concentration in the bulk liquid (mg O_2 L^{-1}).

KL_a: Oxygen coefficient transfer (0.48 h^{-1})

SO_{2_S}: Saturation oxygen concentration (mg O_2 L^{-1}).

O_{2_algae}: Oxygen production by algae (mg O_2 L^{-1}).

O_{2_nitr}: Oxygen consumption by nitrifiers (mg O_2 L^{-1}).

O_{2_OHO}: Oxygen consumption for COD oxidation by OHO (mg O_2 L^{-1}).

$O_{2\_resp_{algae}}$: Oxygen respiration by algae (mg O_2 L^{-1}).

$O_{2\_resp_{OHO}}$: Oxygen respiration by OHO (mg O_2 L^{-1}).

$KL_a(SO_{2\_S} - SO_2)$: Oxygen transfer (mg O_2 L^{-1} h^{-1}).

Eq. B.4.1 was solved to determine the oxygen produced by algae. The solution is stated in Eq. B.4.2:

$$O_{2\_algae} = SO_2 + O_{2\_nitr} + O_{2\_OHO} + O_{2\_resp_{algae}} + O_{2\_resp_{OHO}} - KL_a(SO_{2\_S} - SO_2)$$

Eq. A.4.2

B.4.3 Oxygen transfer

The oxygen transfer was calculated from the oxygen data. For each time step the oxygen transfer was calculated, and later summed up for the two aerobic periods (light phase). The first aerobic period started at 0.3 h and finished at 5 h and the second aerobic started at 5.3 h and finished at 11 h. Eq. B.4.3 was used to calculate the oxygen transfer in terms of mg O_2 L^{-1}.

$$Oxygen\ transfer\ (mg\ O_2L^{-1}) = \sum\left(Kl_a\left(SO_{2_s} - \overline{SO_{2_{t=i;t=i+1}}}\right) * (t_{i+1} - t_i)\right)$$

Eq. B.4.3

B.4.4 OHO bacterial respiration

The endogenous bacterial respiration by OHO ($O_{2\_resp_{OHO}}$) was calculated in terms of mass of oxygen utilized per day based on Eq. B.4.4 (Ekama and Wentzel, 2008b). This equation takes into account the endogenous fraction within the OHO biomass and the COD load in the reactor.

$$O_{2\_resp_{OHO}} = S_{COD}\left[(1-f_H)b_H\frac{Y_{Hv}f_{cv}SRT}{(1+b_H SRT)}\right] \quad \text{Eq. B.4.4}$$

Where:

f_{cv}: COD/VSS ratio for activated sludge (mg COD mg VSS^{-1}).

B.4.5 Algal respiration

The algal respiration was calculated taking into account the dark zones in the reactor, as algal respiration is higher at dark conditions. To determine the dark zone (Dz) in the reactor at different SRTs, the incident light intensity was measured on 12 points outside the reactor wall (Figure B.4.1a). The average incident light intensity was 25.9 µmol m^{-2} s^{-1}. To calculate the penetration of light a different points inside the reactor the Lambert-Beer equation was used. In order to simplify the calculation, it was assumed that the reactor had a rectangular shape with a light path of 0.075 m, which corresponds to the radius of the circular reactor (Figure B.4.1b). This calculation was done for each period. In period 1, the light penetration was just up to 1 cm. For the periods 2A, 2B and 2C, the light reach up to 2 to 4 cm inside the reactor (Figure B.4.2).

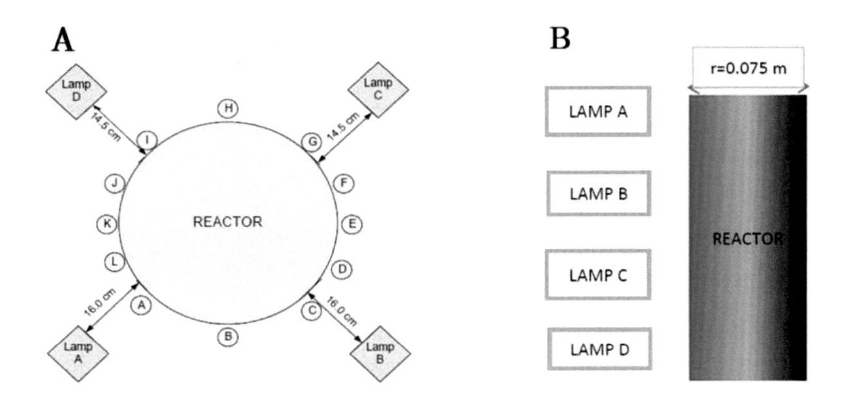

Figure B.4.1. (A) Incident light over the photobioreactor and location of the measuring points and (B) rectangular representation of the photobioreactor for simplification of the light measurement

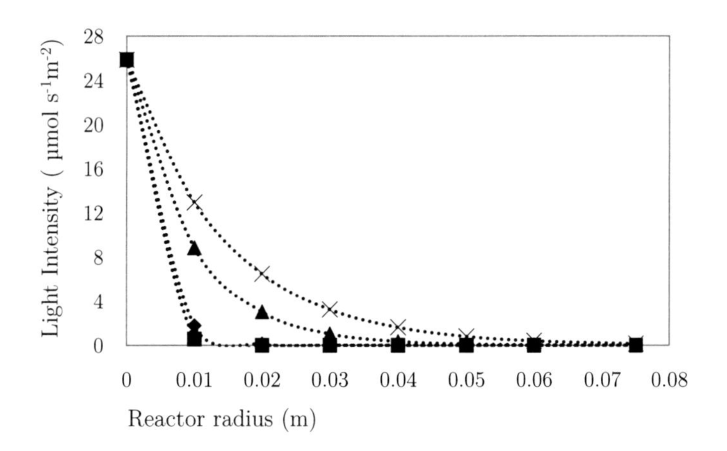

Figure B.4.2. Light penetration from the surface to the centre of the photobioreactor estimated using the Lambert-Beer equation.

Based on this data, the dark zones inside the reactor were estimated. The values were 0.9, 0.78, 0.36 and 0.17% for periods 1, 2A, 2B and 2C.

Furthermore, with the value of the dark zones, the dark respiration coefficient of 0.1 d^{-1} (Zambrano et al., 2016) and the algal biomass (calculated in the biomass characterization), the algal respiration was calculated using Eq. B.4.5.:

$$O_{2\_resp_{algae}} = X_{VSS\_\mathrm{algae}} \, f_{cv,algae} \, Dz \, b_{algae} \, Cycle\ time \qquad \text{Eq. B.4.5}$$

Where:

$O_{2\_resp_{algae}}$: Oxygen respire by algae (mg O_2 L^{-1}).

X_{VSS_algae}: Algal biomass concentration (mg VSS L^{-1}).

$f_{cv,algae}$: COD/VSS ratio for algal biomass (0.953 mg COD mg VSS^{-1}).

Dz: Dark zone fraction within the reactor.

b_{algae}: Dark respiration coefficient for algal biomass (0.1 d^{-1})

B.4.6 Oxygen consumption by nitrifiers and COD oxidation.

The oxygen consumed by nitrification was calculated based on the nitrate formation and the O_2 requirements of 4.57 mg O_2 mg NH_4^+- N^{-1} nitrified to nitrate, as defined by Ekama and Wentzel (2008b). The maximum nitrate formation in a determined period of time and Eq. B.1.3 was used to determine the nitrification.

The oxygen requirement for COD oxidation (O_{2_OHO}) was determined based on the total amount of COD oxidised, which corresponds to the total COD removed minus the COD required for denitrification. Eq. A.4.6 utilized was proposed by Ekama and Wentzel (2008b):

$$O_{2\_OHO} = S_{COD_{oxi}}(1 - f_{cv}Y_{Hv}) \qquad \text{Eq. B.4.6}$$

Where:

O_{2_OHO}: Oxygen used for COD oxidation (mg O_2 L^{-1}).

$S_{COD_{oxi}}$: COD concentration oxidized (mg COD L^{-1}).

C

Appendix C

C.1. MODEL INPUTS, PROCESSES, RATES AND STOICHIOMETRY

Table C.1 List of variables used as input for the model (R1 and R2 at day 49 of phase 2)

Symbol	State variable	Unit	R1	R2
L	Effective light path	m	0.08	0.08
I_0	Incident irradiance	μmol photon m^{-2} s^{-1}	84	84
S_I	Soluble inert organics	g COD m^{-3}	10	10
S_{NH_4}	Ammonium	g N m^{-3}	88.69	87.20
S_{NO_2}	Nitrite nitrogen	g N m^{-3}	1.63	5.5
S_{NO_3}	Nitrate nitrogen	g N m^{-3}	0.21	0.93
S_{N_2}	Nitrogen gas	g N m^{-3}	0	0
S_{O_2}	Dissolved oxygen	g O_2 m^{-3}	2.33	1.45
S_S	Readily biodegradable substrate	g COD m^{-3}	746	646
X_{AOB}	AOB biomass	g COD m^{-3}	300	265
X_{NOB}	NOB biomass	g COD m^{-3}	5	8
X_I	Inert particulate organics	g COD m^{-3}	50	66
X_H	Heterotrophic biomass	g COD m^{-3}	300	396
X_P	Phototrophic biomass	g COD m^{-3}	1220	1597
X_S	Slowly biodegradable substrate	g COD m^{-3}	748	765
X_{STO}	Organics stored by heterotrophs	g COD m^{-3}	50	50

Table C.2 List of coefficients for model elaboration and calibration.

Symbol	Model parameter	Value	Calibrated value	Unit	Reference
$\mu_{max,AOB}$	Maximum specific growth rate of AOB	0.9		d^{-1}	Kaelin et al. (2009)
$\mu_{max,H}$	Maximum specific growth rate of heterotrophs	2		d^{-1}	Henze et al. (2000)
$\mu_{max,NOB}$	Maximum specific growth rate of NOB	0.65		d^{-1}	Kaelin et al. (2009)
$\mu_{max,P}$	Maximum specific growth rate of phototrophs	0.96	0.85	d^{-1}	Sasi et al. (2011)
b_{AOB}	Respiration rate constant for AOB	0.061		d^{-1}	Iacopozzi et al. (2007)
$b_{AOB,NOx}$	Anoxic respiration rate constant for AOB	0.05		d^{-1}	Henze et al. (2000)
b_{NOB}	Respiration rate constant for NOB	0.061		d^{-1}	Iacopozzi et al. (2007)
$b_{NOB,NOx}$	Anoxic respiration rate constant for NOB	0.05		d^{-1}	Henze et al. (2000)
b_H	Aerobic endogenous respiration rate for heterotrophs	0.1		d^{-1}	Iacopozzi et al. (2007)
b_{STO,O_2}	Aerobic respiration rate for X_{STO}	0.2		d^{-1}	Henze et al. (2000)
b_P	Respiration rate constant for phototrophs	0.09		d^{-1}	Wolf et al. (2007)
f_{SI}	Production of S_I in hydrolysis	0		$(\mathrm{g\,COD}_{S_I})^{-1}$ $(\mathrm{g\,COD}_{X_S})$	Henze et al. (2000)
f_{X_I}	Production of X_I in endogenous respiration	0.2		$\mathrm{g\,COD}_{X_I}$ $(\mathrm{g\,COD}_{X_{BM}})^{-1}$	Henze et al. (2000)

Symbol	Model parameter	Value	Calibrated value	Unit	Reference
I_s	Light intensity saturation	13	4	μmol photon m^{-2} s^{-1}	Martínez et al. (1991)
$i_{N,BM}$	N content of bacterial biomass (X_H, X_{AOB}, X_{NOB})	0.07		g N $(g\ COD_{X_{BM}})^{-1}$	Henze et al. (2000)
$i_{N,P}$	N content of algal biomass (X_P)	0.0657		g N $(g\ COD_{X_P})^{-1}$	Calculated in this work
i_{N,S_I}	N content of S_I	0.01		g N $(g\ COD_{S_I})^{-1}$	Henze et al. (2000)
i_{N,S_S}	N content of S_S	0.03		g N $(g\ COD_{S_S})^{-1}$	Henze et al. (2000)
i_{N,X_I}	N content of X_I	0.02		g N $(g\ COD_{X_I})^{-1}$	Henze et al. (2000)
i_{N,X_S}	N content of X_S	0.04		g N $(g\ COD_{X_S})^{-1}$	Henze et al. (2000)
k	Light extinction coefficient	0.0748		m2/g TSS	Calculated in this work
k_H	Hydrolysis rate constant	3		$g\ COD_{X_S}$ $(g\ COD_{X_H})^{-1} d^{-}$	Henze et al. (2000)
k_{STO}	Storage rate constant	5		$g\ COD_{S_S}$ $(g\ COD_{X_H})^{-1} d^{-}$	Henze et al. (2000)
K_X	Hydrolysis saturation constant	1		$g\ COD_{X_S}$ $(g\ COD_{X_H})^{-1}$	Henze et al. (2000)
K_{STO}	Saturation constant for X_{STO}	1		$g\ COD_{X_{STO}}$ $(g\ COD_{X_H})^{-1}$	Henze et al. (2000)
K_{I,NH_4}	Ammonia inhibition of nitrite oxidation	5		g NH_4^+ N m^{-3}	Iacopozzi et al. (2007)
K_{I,O_2}	Oxygen inhibition for heterotrophs	0.2		g O2 m^{-3}	Kaelin et al. (2009)
$K_{NH_4,AOB}$	Saturation constant for S_{NH_4} for AOB	2.4		g N m^{-3}	Wiesmann (1994)

Symbol	Model parameter	Value	Calibrated value	Unit	Reference
$K_{NH_4,H}$	Saturation constant for $S_{NH_4}$for heterotrophs	0.01		g N m^{-3}	Henze et al. (2000)
$K_{NH_4,P}$	Saturation constant for S_{NH_4} for phototrophs	0.017	0.00021	g NH_3 m^{-3}	Wolf et al. (2007)
$K_{NO_2,,AOB}$	Saturation constant for S_{NO_2} for AOB	0.28		g NO_2^--N m^{-3}	Manser et al. (2005)
$K_{NO_2,,H}$	Saturation constant for S_{NO_2} for heterotrophs	0.5		g NO_3^--N m^{-3}	Henze et al. (2000)
$K_{NO_2,,NOB}$	Saturation constant for S_{NO_2} for NOB	0.28		g NO_2^--N m^{-3}	Manser et al. (2005)
$K_{NO_3,,H}$	Saturation constant for S_{NO_3} for heterotrophs	0.5		g NO_3^--N m^{-3}	Henze et al. (2000)
$K_{NO_3,,NOB}$	Saturation constant for S_{NO_3} for NOB	0.28		g NO_2^--N m^{-3}	Manser et al. (2005)
$K_{O_2,AOB}$	Saturation constant for S_{O_2} for AOB	0.79		g O_2 m^{-3}	Manser et al. (2005)
$K_{O_2,H}$	Saturation constant for S_{O_2} for heterotrophs	0.2		g O_2 m^{-3}	Henze et al. (2000)
$K_{O_2,NOB}$	Saturation constant for S_{O_2} for NOB	0.47		g O_2 m^{-3}	Manser et al. (2005)
$K_{S,H}$	Saturation constant for S_S for heterotrophs	2		g COD_{S_S} m^{-3}	Henze et al. (2000)
η_H	Anoxic reduction factor for heterotrophs	0.6		-	Henze et al. (2000)
Y_{AOB}	Aerobic yield of X_{AOB}	0.2		g $COD_{X_{AOB}}$ (g $COD_{S_{NH_4}})^{-1}$	Sin et al. (2008)
Y_{H,O_2}	Aerobic yield of X_H	0.63		g COD_{X_H} (g $COD_{X_{STO}})^{-1}$	Henze et al. (2000)

Symbol	Model parameter	Value	Calibrated value	Unit	Reference
Y_{H,NO_x}	Anoxic yield of X_H	0.54		g COD_{X_H} (g $COD_{X_{STO}}$)$^{-1}$	Henze et al. (2000)
Y_{NOB}	Aerobic yield of X_{NOB}	0.05		g $COD_{X_{NOB}}$ (g $COD_{S_{NO_2}}$)$^{-1}$	Sin et al. (2008)
Y_{STO,O_2}	Aerobic yield of X_{STO} in of X_H	0.85		g COD_{X_H} (g COD_{S_S})$^{-1}$	Henze et al. (2000)
Y_{STO,NO_2}	Anoxic yield of X_{STO} in of X_H on nitrite	0.8		g COD_{X_H} (g COD_{S_S})$^{-1}$	Henze et al. (2000)
Y_{STO,NO_3}	Anoxic yield of X_{STO} in of X_H on nitrate	0.8		g COD_{X_H} (g COD_{S_S})$^{-1}$	Henze et al. (2000)

Table C.3 Processes and rates of the proposed model.

	Process	Process rate equation	Reference
1	Hydrolysis	$k_H \frac{X_S/X_H}{K_X + (X_S/X_H)} X_H$	Original ASM3
Heterotrophic organisms (aerobic and denitrifying activity)			
2	Aerobic Storage	$k_{STO} \frac{S_{O_2}}{K_{O_2,H} + S_{O_2}} \frac{S_S}{K_{S,H} + S_S} X_H$	Original ASM3
3	Anoxic Storage on nitrite	$k_{STO}\, \eta_H \frac{K_{I,O_2}}{K_{I,O_2} + S_{O_2}} \frac{S_S}{K_{S,H} + S_S} \frac{S_{NO_2}}{K_{NO_2,H} + S_{NO_2}} X_H$	Modified ASM3
4	Anoxic Storage on nitrate	$k_{STO}\, \eta_H \frac{K_{I,O_2}}{K_{I,O_2} + S_{O_2}} \frac{S_S}{K_{S,H} + S_S} \frac{S_{NO_3}}{K_{NO_3,H} + S_{NO_3}} X_H$	Modified ASM3
5	Aerobic Growth	$\mu_{max,H} \frac{S_{O_2}}{K_{O_2,H} + S_{O_2}} \frac{S_{NH_4}}{K_{NH_4,H} + S_{NH_4}} \frac{S_S}{K_{S,H} + S_S} X_H$	Original ASM3
6	Anoxic Growth on nitrite	$\mu_{max,H}\, \eta_H \frac{K_{I,O_2}}{K_{I,O_2} + S_{O_2}} \frac{S_{NH_4}}{K_{NH_4,H} + S_{NH_4}} \frac{X_{STO}/X_H}{K_{STO} + (X_{STO}/X_H)} \frac{S_{NO_2}}{K_{NO_2,H} + S_{NO_2}} X_H$	Modified ASM3
7	Anoxic Growth on nitrate	$\mu_{max,H}\, \eta_H \frac{K_{I,O_2}}{K_{I,O_2} + S_{O_2}} \frac{S_{NH_4}}{K_{NH_4,H} + S_{NH_4}} \frac{X_{STO}/X_H}{K_{STO} + (X_{STO}/X_H)} \frac{S_{NO_3}}{K_{NO_3,H} + S_{NO_3}} X_H$	Modified ASM3
8	Aerobic End. Respiration of X_{STO}	$b_{STO,O_2} \frac{S_{O_2}}{K_{O_2,H} + S_{O_2}} X_{STO}$	Original ASM3
9	Anoxic End. Respiration of X_{STO} on nitrite	$b_{STO,O_2}\, \eta_H \frac{K_{I,O_2}}{K_{I,O_2} + S_{O_2}} \frac{S_{NO_2}}{K_{NO_2,H} + S_{NO_2}} X_{STO}$	Modified ASM3
10	Anoxic End. Respiration of X_{STO} on nitrate	$b_{STO,O_2}\, \eta_H \frac{K_{I,O_2}}{K_{I,O_2} + S_{O_2}} \frac{S_{NO_3}}{K_{NO_3,H} + S_{NO_3}} X_{STO}$	Modified ASM3
11	Aerobic End. Respiration	$b_H \frac{S_{O_2}}{K_{O_2,H} + S_{O_2}} X_H$	Original ASM3
12	Anoxic End. Respiration on nitrite	$b_H\, \eta_H \frac{K_{I,O_2}}{K_{I,O_2} + S_{O_2}} \frac{S_{NO_2}}{K_{NO_2,H} + S_{NO_2}} X_H$	Modified ASM3

13	Anoxic End. Respiration on nitrate	$b_H\, \eta_H \frac{K_{I,O_2}}{K_{I,O_2} + S_{O_2}} \frac{S_{NO_3}}{K_{NO_3,H} + S_{NO_3}} X_H$	Modified ASM3
Autotrophic organisms (nitrifying activity)			
14	Aerobic Growth (AOB)	$\mu_{max,AOB} \frac{S_{O_2}}{K_{O_2,AOB} + S_{O_2}} \frac{S_{NH_4}}{K_{NH_4,AOB} + S_{NH_4}} X_{AOB}$	Modified ASM3
15	Aerobic End. Respiration (AOB)	$b_{AOB} \frac{S_{O_2}}{K_{O_2,AOB} + S_{O_2}} X_{AOB}$	Modified ASM3
16	Anoxic End. Respiration (AOB)	$b_{AOB,NOx} \frac{K_{O_2,AOB}}{K_{O_2,AOB} + S_{O_2}} \frac{S_{NO_2}}{K_{NO_2,AOB} + S_{NO_2}} X_{AOB}$	Modified ASM3
17	Aerobic Growth (NOB)	$\mu_{max,NOB} \frac{S_{O_2}}{K_{O_2,NOB} + S_{O_2}} \frac{S_{NO_2}}{K_{NO_2,NOB} + S_{NO_2}} \frac{K_{I,NH_4}}{K_{I,NH_4} + S_{NH_4}} X_{NOB}$	Modified ASM3
18	Aerobic End. Respiration (NOB)	$b_{NOB} \frac{S_{O_2}}{K_{O_2,NOB} + S_{O_2}} X_{NOB}$	Modified ASM3
19	Anoxic End. Respiration (NOB)	$b_{NOB,NOx} \frac{K_{O_2,NOB}}{K_{O_2,NOB} + S_{O_2}} \frac{S_{NO_3}}{K_{NO_3,NOB} + S_{NO_3}} X_{NOB}$	Modified ASM3
Phototrophic organisms			
20	Phototrophic Growth considering NH_4^+ and Light intensity	$\mu_{max,P} \frac{S_{NH_4}}{K_{NH_4,P} + S_{NH_4}} \left\{1 - \exp\left(\frac{-I_o[1 - \exp(-k\, X_T\, L)]}{k\, X_T\, L\, I_s}\right)\right\} X_P$	This work
21	Phototrophic End. Respiration	$b_P\, X_P$	This work

Table C.4 Stoichiometric matrix of the proposed model.

Processes / Variables	X_H	X_{AOB}	X_{NOB}	X_P	X_I	X_S	X_{STO}	S_{O_2}	S_{NH_4}	S_{NO_3}	S_{NO_2}	S_{N_2}	S_S	S_I
1 Hydrolysis						-1			$-i_{N,S_S}(1-f_{SI})-(f_{SI}i_{N,S_I})$ $+i_{N,X_S}$				$1-f_{SI}$	f_{SI}
Heterotrophic organisms (denitrifiers)														
2 Aerobic Storage							Y_{STO,O_2}	$Y_{STO,O_2}-1$	i_{N,S_S}				-1	
3 Anoxic Storage on nitrite							Y_{STO,NO_2}		i_{N,S_S}		$-\frac{(1-Y_{STO,NO_2})}{1.71}$	$\frac{(1-Y_{STO,NO_2})}{1.71}$	-1	
4 Anoxic Storage on nitrate							Y_{STO,NO_3}		i_{N,S_S}	$-\frac{(1-Y_{STO,NO_3})}{1.14}$	$\frac{(1-Y_{STO,NO_3})}{1.14}$		-1	
5 Aerobic Growth	1							$-\frac{1-Y_{H,O_2}}{Y_{H,O_2}}$	$-i_{N,BM}$				$-\frac{1}{Y_{H,O_2}}$	
6 Anoxic Growth on nitrite	1								$-i_{N,BM}$		$-\frac{1-Y_{H,NOx}}{1.71\cdot Y_{H,NOx}}$	$\frac{1-Y_{H,NOx}}{1.71\cdot Y_{H,NOx}}$	$-\frac{1}{Y_{H,NOx}}$	
7 Anoxic Growth on nitrate	1								$-i_{N,BM}$	$-\frac{1-Y_{H,NOx}}{1.14\cdot Y_{H,NOx}}$	$\frac{1-Y_{H,NOx}}{1.14\cdot Y_{H,NOx}}$		$-\frac{1}{Y_{H,NOx}}$	
8 Aerobic End. Respiration of X_{STO}							-1	-1						
9 Anoxic End. Respiration of X_{STO} on nitrite							-1				$-\frac{1}{1.72}$	$\frac{1}{1.72}$		
10 Anoxic End. Respiration of X_{STO} on nitrate							-1			$-\frac{1}{1.14}$	$\frac{1}{1.14}$			
11 Aerobic End. Respiration	-1				f_{X_I}			$-(1-f_{X_I})$	$i_{N,BM}-f_{X_I}i_{N,X_I}$					
12 Anoxic End. Respiration on nitrite	-1				f_{X_I}				$i_{N,BM}-f_{X_I}i_{N,X_I}$		$-\frac{1-f_{X_I}}{1.71}$	$\frac{1-f_{X_I}}{1.71}$		
13 Anoxic End. Respiration on nitrate	-1				f_{X_I}				$i_{N,BM}-f_{X_I}i_{N,X_I}$	$-\frac{1-f_{X_I}}{1.14}$	$\frac{1-f_{X_I}}{1.14}$			

Processes / Variables	X_H	X_{AOB}	X_{NOB}	X_P	X_I	X_S	X_{STO}	S_{O_2}	S_{NH_4}	S_{NO_3}	S_{NO_2}	S_{N_2}	S_S	S_I
Autotrophic organisms (nitrifiers)														
14 Aerobic Growth (AOB)		1						$-\frac{3.43 - Y_{AOB}}{Y_{AOB}}$	$-i_{N,BM} - \frac{1}{Y_{AOB}}$		$\frac{1}{Y_{AOB}}$			
15 Aerobic End. Respiration (AOB)		−1			f_{X_I}			$-(1 - f_{X_I})$	$i_{N,BM} - f_{X_I} i_{N,X_I}$					
16 Anoxic End. Respiration (AOB)		−1			f_{X_I}				$i_{N,BM} - f_{X_I} i_{N,X_I}$		$-\frac{1 - f_{X_I}}{2.86}$	$\frac{1 - f_{X_I}}{2.86}$		
17 Aerobic Growth (NOB)			1					$-\frac{1.14 - Y_{NOB}}{Y_{NOB}}$	$-i_{N,BM}$	$\frac{1}{Y_{NOB}}$	$-\frac{1}{Y_{NOB}}$			
18 Aerobic End. Respiration (NOB)			−1		f_{X_I}			$-(1 - f_{X_I})$	$i_{N,BM} - f_{X_I} i_{N,X_I}$					
19 Anoxic End. Respiration (NOB)			−1		f_{X_I}				$i_{N,BM} - f_{X_I} i_{N,X_I}$	$-\frac{1 - f_{X_I}}{2.86}$		$\frac{1 - f_{X_I}}{2.86}$		
Phototrophic organisms														
20 Phototrophic Growth considering NH_4^+ and Light intensity				1				0.96	$-i_{N,P}$					
21 Phototrophic End. Respiration				−1	f_{X_I}			−0.96	$i_{N,P} - f_{X_I} i_{N,X_I}$					

D

Appendix D

D.1 CALIBRATED AND THEORETICAL PARAMETERS OF THE MICROALGAL-BACTERIAL MODEL

Table D.1 Literature and calibrated values of the microalgal-bacterial model

Symbol	Model parameter	Literature value	Calibrated value	Unit	Reference
$\mu_{max,AOB}$	Maximum specific growth rate of AOB		1.1±0.02	d^{-1}	
$\mu_{max,H}$	Maximum specific growth rate of heterotrophs		5.5 ±0.02	d^{-1}	
$\mu_{max,NOB}$	Maximum specific growth rate of NOB		1.3 ±0.01	d^{-1}	
$\mu_{max,P}$	Maximum specific growth rate of phototrophs		2.0 ±0.05	d^{-1}	
b_{AOB}	Respiration rate constant for AOB	0.061		d^{-1}	Iacopozzi et al. (2007)
$b_{AOB,NOx}$	Anoxic respiration rate constant for AOB	0.05		d^{-1}	Henze et al. (2000)
b_{NOB}	Respiration rate constant for NOB	0.061		d^{-1}	Iacopozzi et al. (2007)
$b_{NOB,NOx}$	Anoxic respiration rate constant for NOB	0.05		d^{-1}	Henze et al. (2000)
b_H	Aerobic endogenous respiration rate for heterotrophs	0.1		d^{-1}	Iacopozzi et al. (2007)
b_{STO,O_2}	Aerobic respiration rate for X_{STO}	0.2		d^{-1}	Henze et al. (2000)
b_P	Respiration rate constant for phototrophs	0.09		d^{-1}	Wolf et al. (2007)
f_{SI}	Production of S_I in hydrolysis	0		$(g\ COD_{S_I})^{-}$ $(g\ COD_{X_S})$	Henze et al. (2000)

f_{X_I}	Production of X_I in endogenous respiration	0.2		$g\ COD_{X_I}$ $(g\ COD_{X_{BM}})$	Henze et al. (2000)
I_s	Light intensity saturation		35.0±0.4	mol photon $m^{-2}\ s^{-1}$	
$i_{N,BM}$	N content of bacterial biomass (X_H, X_{AOB}, X_{NOB})	0.07		g N $(g\ COD_{X_{BM}})$	Henze et al. (2000)
$i_{N,P}$	N content of algal biomass (X_P)	0.0657		g N $(g\ COD_{X_P})^{-}$	Calculated in this work
i_{N,S_I}	N content of S_I	0.01		g N $(g\ COD_{S_I})^{-1}$	Henze et al. (2000)
i_{N,S_S}	N content of S_S	0.03		g N $(g\ COD_{S_S})^{-1}$	Henze et al. (2000)
i_{N,X_I}	N content of X_I	0.02		g N $(g\ COD_{X_I})^{-1}$	Henze et al. (2000)
i_{N,X_S}	N content of X_S	0.04		g N $(g\ COD_{X_S})^{-}$	Henze et al. (2000)
k	Light extinction coefficient		0.019 ±0.003	m^2/g TSS	Calculated in this work
k_H	Hydrolysis rate constant	3		$g\ COD_{X_S}$ $(g\ COD_{X_H})^{-}$	Henze et al. (2000)
k_{STO}	Storage rate constant		0.88±0.03	$g\ COD_{S_S}$ $(g\ COD_{X_H})^{-}$	Henze et al. (2000)
K_X	Hydrolysis saturation constant	1		$g\ COD_{X_S}$ $(g\ COD_{X_H})^{-}$	Henze et al. (2000)
K_{STO}	Saturation constant for X_{STO}	1		$g\ COD_{X_{STO}}$ $(g\ COD_{X_H})^{-}$	Henze et al. (2000)
K_{I,NH_4}	Ammonia inhibition of nitrite oxidation	5		g NH_4^+ N m^{-3}	Iacopozzi et al. (2007)

K_{I,O_2}	Oxygen inhibition for heterotrophs	0.2		g O_2 m^{-3}	Kaelin et al. (2009)
$K_{NH_4,AOB}$	Saturation constant for S_{NH_4} for AOB		0.13±0.02	g N m^{-3}	
$K_{NH_4,H}$	Saturation constant for S_{NH_4} for heterotrophs	0.01		g N m^{-3}	Henze et al. (2000)
$K_{NH_4,P}$	Saturation constant for S_{NH_4} for phototrophs		0.01±0.00	g NH_3 m^{-3}	
$K_{NO_2,,AOB}$	Saturation constant for S_{NO_2} for AOB	0.28		g NO_2^- N m^{-3}	Manser et al. (2005)
$K_{NO_2,,H}$	Saturation constant for S_{NO_2} for heterotrophs	0.5		g NO_3 N m^{-3}	Henze et al. (2000)
$K_{NO_2,,NOB}$	Saturation constant for S_{NO_2} for NOB	0.28		g NO_2^- N m^{-3}	Manser et al. (2005)
$K_{NO_3,,H}$	Saturation constant for S_{NO_3} for heterotrophs	0.5		g NO_3^- N m^{-3}	Henze et al. (2000)
$K_{NO_3,,NOB}$	Saturation constant for S_{NO_3} for NOB	0.28		g NO_2^- N m^{-3}	Manser et al. (2005)
$K_{O_2,AOB}$	Saturation constant for S_{O_2} for AOB		0.75±0.01	g O_2 m^{-3}	
$K_{O_2,H}$	Saturation constant for S_{O_2} for heterotrophs	0.2		g O_2 m^{-3}	Henze et al. (2000)
$K_{O_2,NOB}$	Saturation constant for S_{O_2} for NOB	0.47		g O_2 m^{-3}	Manser et al. (2005)
$K_{S,H}$	Saturation constant for S_S for heterotrophs	2		g COD_{S_S} m^{-3}	Henze et al. (2000)
η_H	Anoxic reduction factor for heterotrophs	0.6		-	Henze et al. (2000)
Y_{AOB}	Aerobic yield of X_{AOB}	0.2		g $COD_{X_{AOB}}$ (g $COD_{S_{NH_4}}$	Sin et al. (2008)

Y_{H,O_2}	Aerobic yield of X_H	0.63	g COD_{X_H} (g $COD_{X_{STO}}$)	Henze et al. (2000)
Y_{H,NO_x}	Anoxic yield of X_H	0.54	g COD_{X_H} (g $COD_{X_{STO}}$)	Henze et al. (2000)
Y_{NOB}	Aerobic yield of X_{NOB}	0.05	g $COD_{X_{NOB}}$ (g $COD_{S_{NO_2}}$)	Sin et al. (2008)
Y_{STO,O_2}	Aerobic yield of X_{STO} in of X_H	0.85	g COD_{X_H} (g $COD_{S_S})^{-1}$	Henze et al. (2000)
Y_{STO,NO_2}	Anoxic yield of X_{STO} in of X_H on nitrite	0.8	g COD_{X_H} (g $COD_{S_S})^{-1}$	Henze et al. (2000)
Y_{STO,NO_3}	Anoxic yield of X_{STO} in of X_H on nitrate	0.8	g COD_{X_H} (g $COD_{S_S})^{-1}$	Henze et al. (2000)

D.2 CALIBRATION AND VALIDATION OF THE MICROALGAL-BACTERIAL MODEL IN BATCH MODE

Table D.2 Calibrated parameters for the algal-bacterial model

Symbol	Model parameter	Value	Unit
I_s	Light intensity saturation	35.0±0.4	μmol photon m^{-2} s^{-1}
k	Light extinction coefficient	0.005±0.001	m^2 $gTSS^{-1}$
$k_{NH4,AOB}$	Half saturation constant for S_{NH_4} for AOB	0.13±0.02	g N m^{-3}
$K_{O2,AOB}$	Half saturation constant for S_{O_2} for AOB	0.75±0.01	g O_2 m^{-3}
$K_{NH4,P}$	Half saturation constant for S_{NH_4} for phototrophs	0.01±0.00	g N m^{-3}
$\mu_{m,AOB}$	Maximum growth rate of AOB	0.45±0.02	d^{-1}
$\mu_{m,NOB}$	Maximum growth rate of NOB	0.31±0.01	d^{-1}
$\mu_{m,P}$	Maximum growth rate of phototrophs	0.85±0.05	d^{-1}
k_{STO}	Storage rate constant	0.88±0.03	g COD_{Ss} g COD_{XH}^{-1} d^{-1}

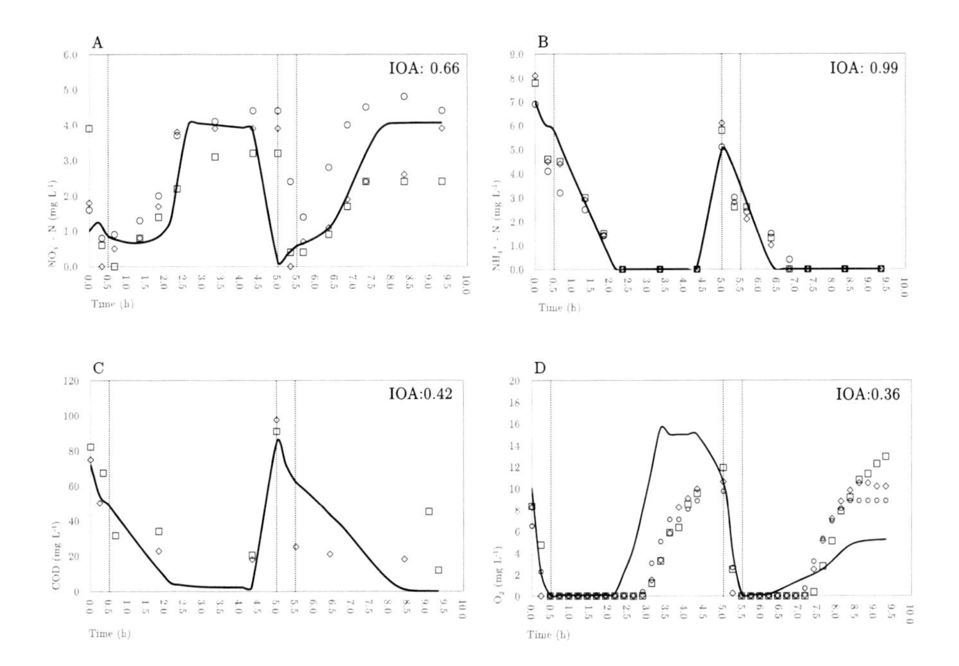

Figure D.1 Modelled and measured data for (A) NO_3^--N, (B) NH_4^+-N, (C) COD and (D) O_2 concentrations during period 1A. Solid line: model data during period 2A; measured data during period 1A in cycles C1 (□), C2 (◇), C3 (○).

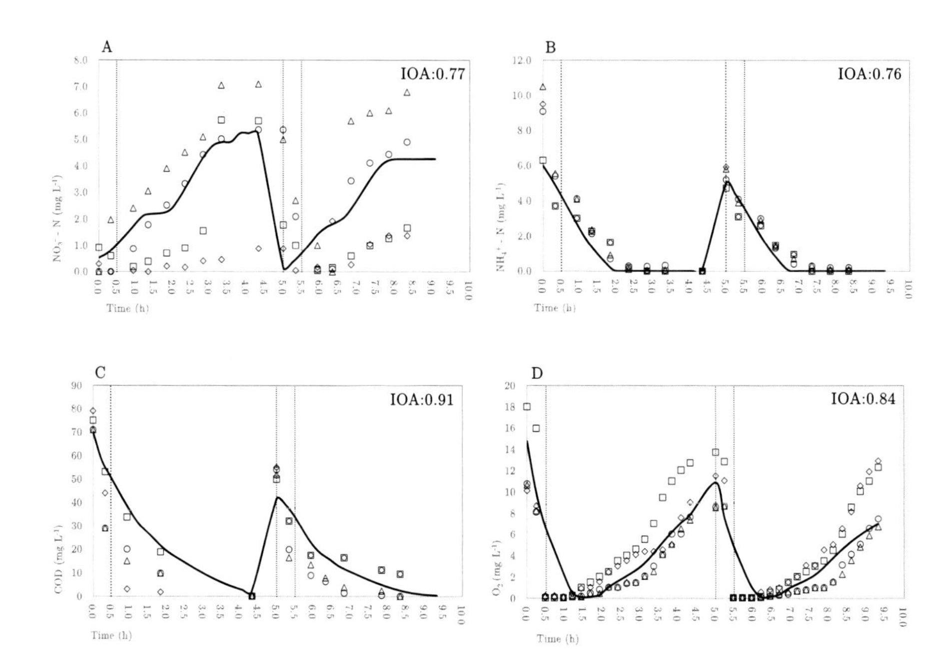

Figure D.2 Modelled and measured data for (A) NO_3^--N, (B) NH_4^+-N, (C) COD and (D) O_2 concentrations during period 2C. Solid line: model data during period 2C; measured data during period 2C in cycles C1 (□), C2 (◇), C3 (○), C4 (△).

D.3 CALIBRATION AND VALIDATION OF THE MICROALGAL-BACTERIAL MODEL IN SEQUENCING BATCH MODE

D.3.1 Biomass production validation

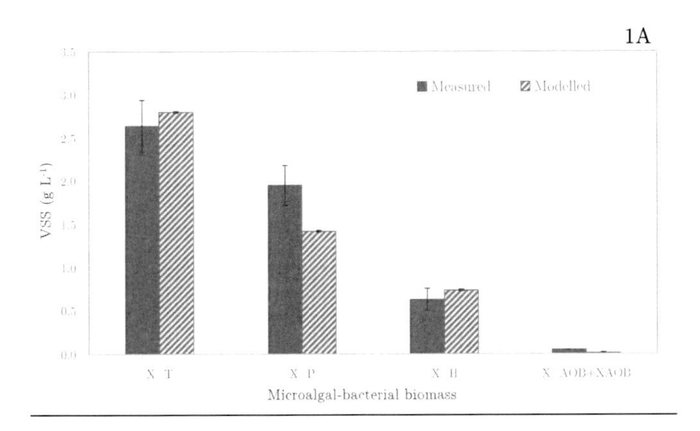

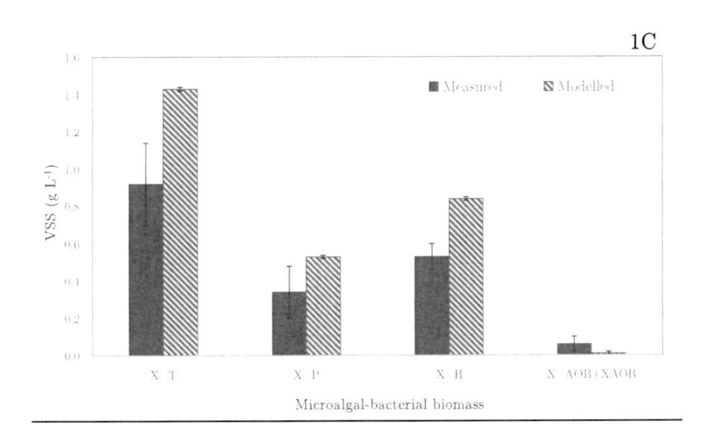

Figure D.3 Biomass modelled in the microalgal-bacterial reactor for period 1A and 1C.

D.3.2 N-compound concentrations measured and modelled in sequencing batch mode

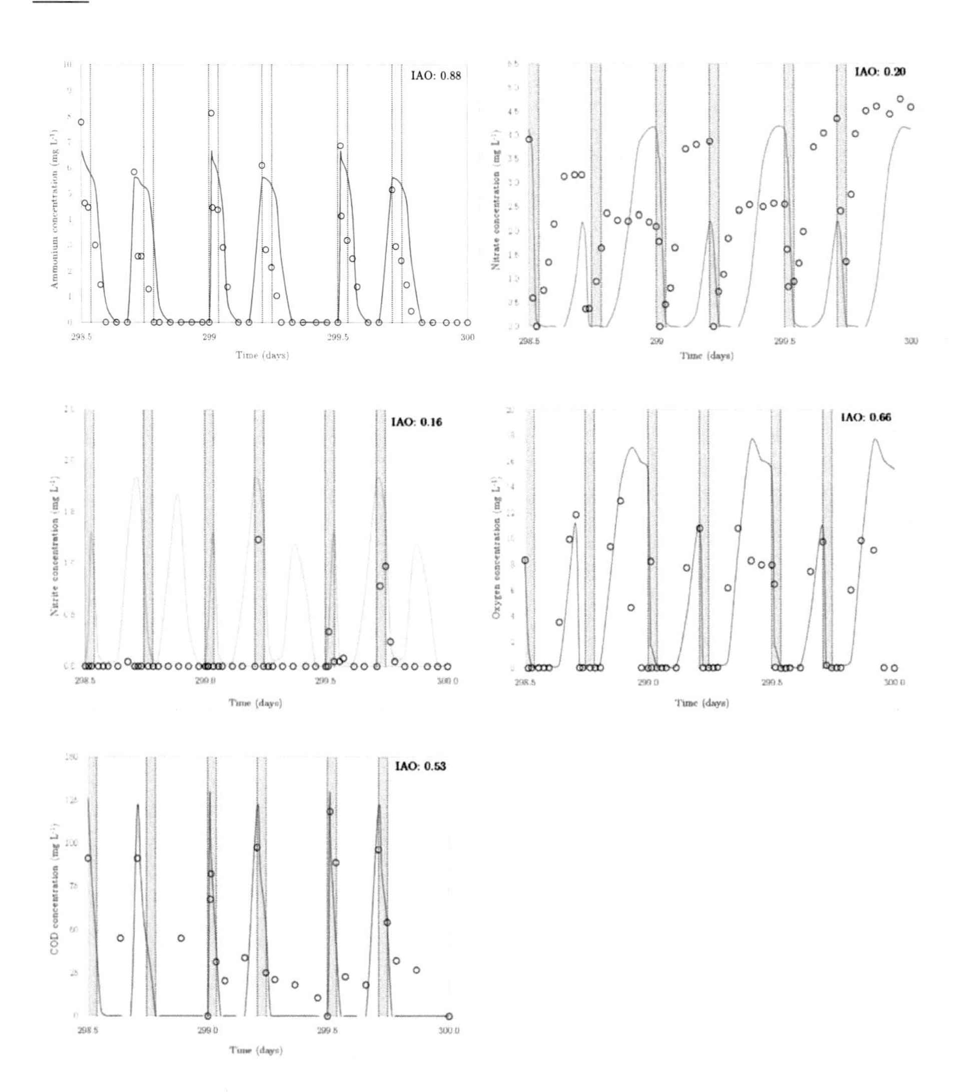

Figure D.4 Modelled and measured data for ammonium, nitrate, nitrite, oxygen and COD during period 1A. Ammonium (▬), nitrate (▬), nitrite concentration (▬), oxygen (▬), COD (▬) and measured data (-o-).

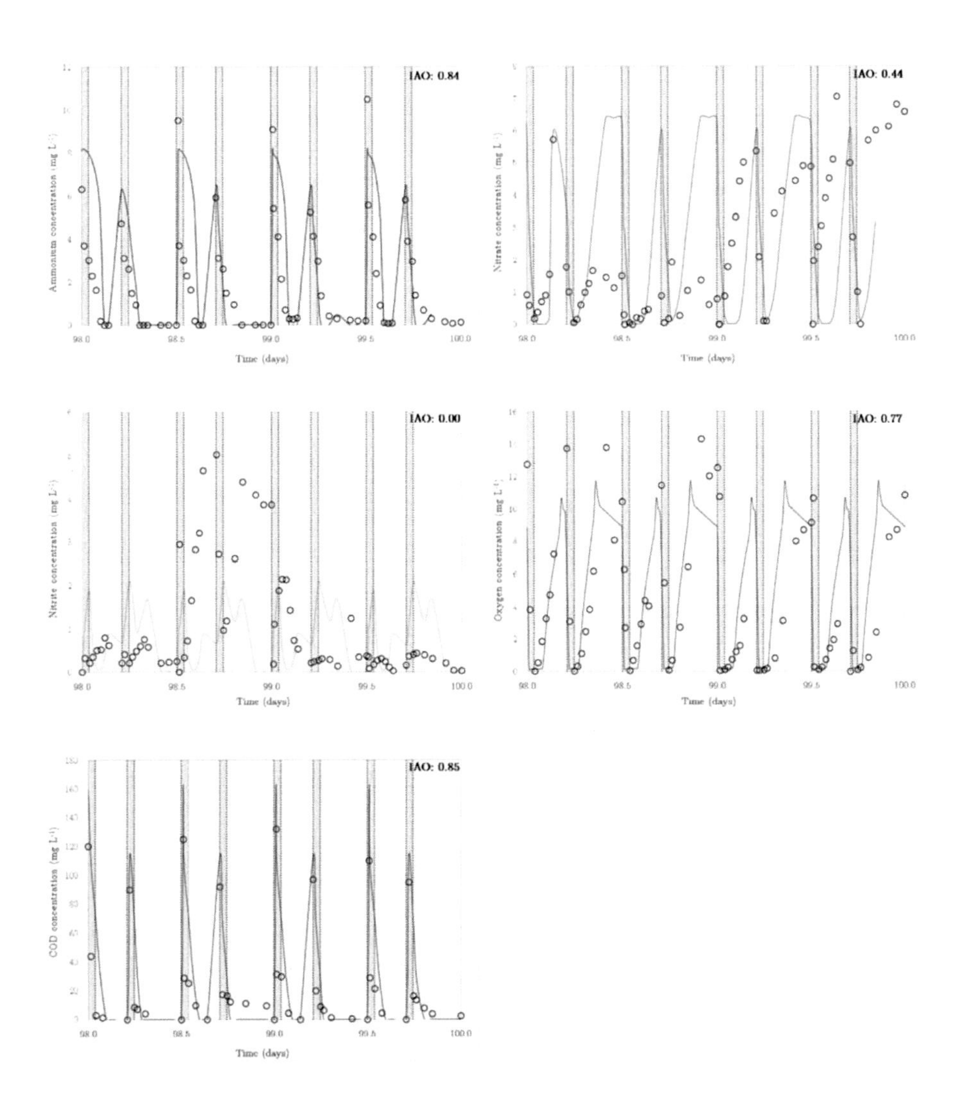

Figure D.5 Modelled and measured data for ammonium, nitrate, nitrite, oxygen and COD during period 1C. Ammonium (▬), nitrate (▬), nitrite concentration (▬), oxygen (▬), COD (▬) and measured data (-o-).

D.3 SRT SCENARIOS

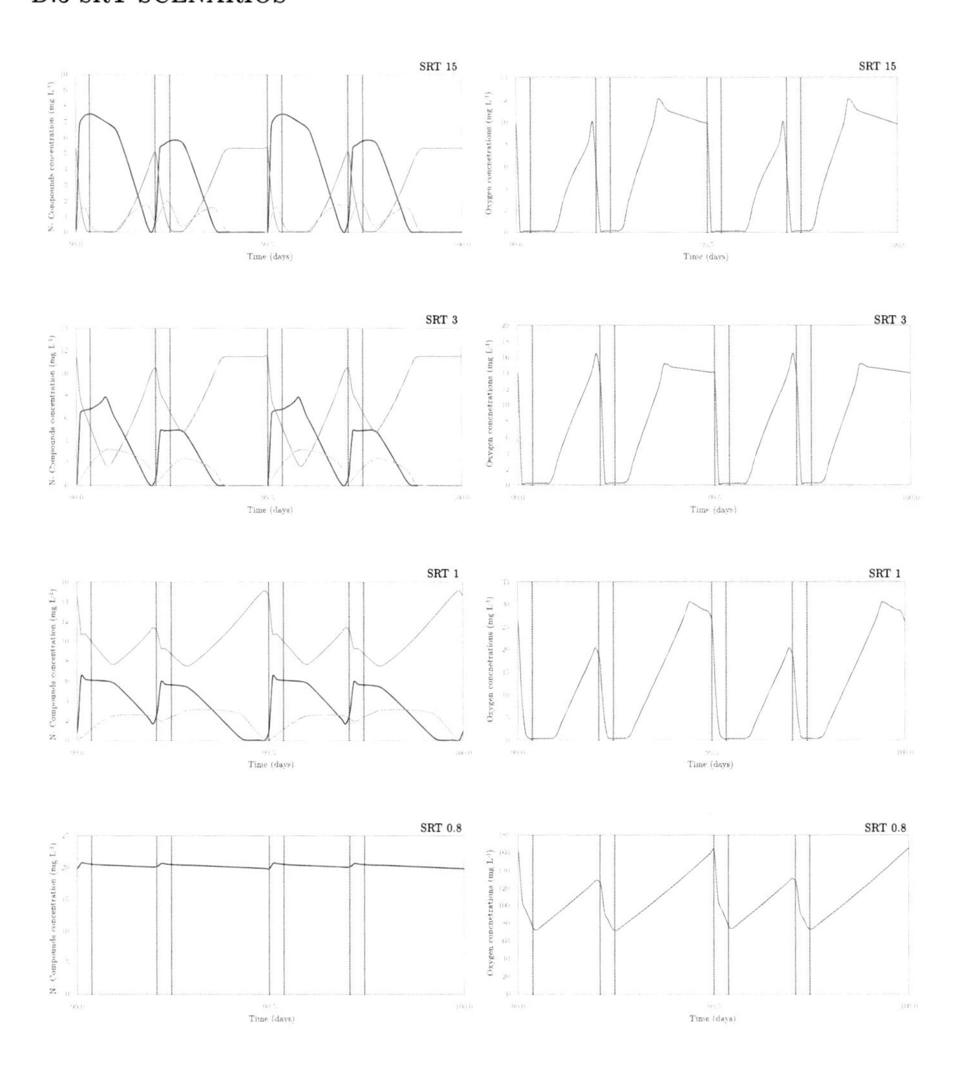

Figure D.7 N-compounds concentration and oxygen concentration for modelled SRTs: 15, 3, 1 and 0.8 days . Ammonium (▬), nitrate (▬), nitrite concentration (▬), oxygen (▬).

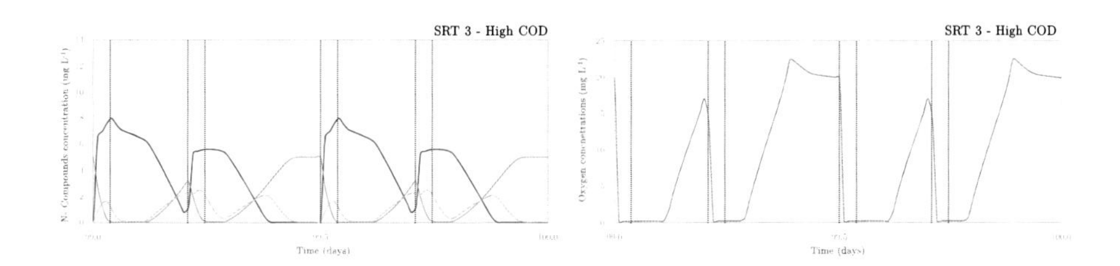

Figure D.8 N-compounds concentration and oxygen concentration for 3 days SRT and COD concentration of 175 mgCOD L^{-1}. Ammonium (▬), nitrate (▬), nitrite concentration (▬), oxygen (▬).

E

Appendix E

Appendix E. New parameters included in the mathematical model

Table E.1 New included parameters in the algal-bacterial model

Symbol	Model parameter	Unit
X_N	Stored nitrogen concentration	g N m^{-3}
$k_{sto,N}$	Storage rate of nitrogen in the algal biomass	g N gCODX_P^{-1} d^{-1}
$K_{STO,N}$	Saturation constant for nitrogen storage	g N m^{-3}
$f_{n,max}$	Maximum fraction of nitrogen stored in the microalgal biomass	gN gCODX_P-1
f_n	Fraction of nitrogen stored in the microalgal biomass	g N gCODX_P-1
$\mu_{m,STO,N}$	Microalgal maximum growth rate on stored nitrogen	d-1
$f_{X_N}^{REG}$	Regulation constant of the microalgal biomass	g N_{STO} g COD_{XP}^{-1}

Table E.2 New processes and rates for the phototrophic organisms included in the mathematical model

	Process	Process rate equation	Reference
Phototrophic organisms			
20	Nitrogen storage	$\frac{S_{NH_4}}{S_{NH_4}+K_{s,NH_4}} k_{sto,N} \frac{f_{n,max}-f_n}{(f_{n,max}-f_n)+K_{STO,N}} X_P$	This work
21	Phototrophic Growth considering NH_4^+ and Light intensity	$\mu_{max,P} \frac{S_{NH_4}}{K_{NH_4,P}+S_{NH_4}} \left\{1-\exp\left(\frac{-I_o[1-\exp(-k\,X_T\,L)]}{k\,X_T\,L\,I_s}\right)\right\} X_P$	This work
22	Phototrophic Growth on NH_4^+ stored and Light intensity	$\mu_{m,STO,N} \frac{S_{NH_4}}{K_{NH_4,P}+S_{NH_4}} \left\{1-\exp\left(\frac{-I_o[1-\exp(-k\,X_T\,L)]}{k\,X_T\,L\,I_s}\right)\right\} \left\{\frac{\frac{X_N}{X_P}}{K_{STO,N,P}+\frac{X_N}{X_P}} \frac{\frac{X_N}{X_P}}{f_{X_N}^{REG}}\right\} X_P$	This work
23	Phototrophic End. Respiration	$b_P\,X_P$	This work

Table E.3 New included processes for the phototrophic organisms in the algal-bacterial model

	Processes / Variables	X_H	X_{AOB}	X_{NOB}	X_P	X_I	X_S	X_{STO}	X_N	S_{O_2}	S_{NH_4}	S_{NO_3}	S_{NO_2}	S_{N_2}	S_S	S_I
Phototrophic organisms																
20	Nitrogen storage								1		−1					
21	Phototrophic Growth considering NH_4^+ and Light intensity				1					0.96	$-i_{N,P}$					
22	Phototrophic Growth on NH_4^+ stored and Light intensity				1				$-i_{N,P}$	0.96						
23	Phototrophic End. Respiration				−1	f_{X_I}				−0.96	$i_{N,P} - f_{X_I} i_{N,X_I}$					

REFERENCES

Abdel-Raouf, N., Al-Homaidan, A.A., Ibraheem, I.B.M., 2012. Microalgae and wastewater treatment. Saudi J. Biol. Sci. 19, 257–275. https://doi.org/10.1016/j.sjbs.2012.04.005

Abed, R.M.M., Zein, B., Al-Thukair, A., de Beer, D., 2007. Phylogenetic diversity and activity of aerobic heterotrophic bacteria from a hypersaline oil-polluted microbial mat. Syst. Appl. Microbiol. 30, 319–330. https://doi.org/10.1016/j.syapm.2006.09.001

Abou-Shanab, R.A.I., Ji, M.-K., Kim, H.-C., Paeng, K.-J., Jeon, B.-H., 2013. Microalgal species growing on piggery wastewater as a valuable candidate for nutrient removal and biodiesel production. J. Environ. Manage. 115, 257–264. https://doi.org/10.1016/j.jenvman.2012.11.022

Alcántara, C., Posadas, E., Guieysse, B., Muñoz, R., 2015. Chapter 29 - Microalgae-based Wastewater Treatment, in: Kim, S.-K. (Ed.), Handbook of Marine Microalgae. Academic Press, Boston, pp. 439–455.

Anthonisen, A.C., Loehr, R.C., Prakasam, T.B.S., Srinath, E.G., 1976. Inhibition of nitrification by ammonia and nitrous Acid. J. Water Pollut. Control Fed. 48, 835–852.

APHA, 2005. Standard Methods for the Examination of Water & Wastewater: Contennial Edition, 21 Har/Cdr. ed. American Technical Publishers.

Arashiro, L.T., Rada-Ariza, A.M., Wang, M., Steen, P. van der, Ergas, S.J., 2016. Modelling shortcut nitrogen removal from wastewater using an algal-bacterial consortium. Water Sci. Technol 75. 782 - 792. https://doi.org/10.2166/wst.2016.561

Aslan, S., Kapdan, I.K., 2006. Batch kinetics of nitrogen and phosphorus removal from synthetic wastewater by algae. Ecol. Eng. 28, 64–70. https://doi.org/10.1016/j.ecoleng.2006.04.003

Azimi, A.A., Horan, N.J., 1991. The influence of reactor mixing characteristics on the rate of nitrification in the activated sludge process. Water Res. 25, 419–423. https://doi.org/10.1016/0043-1354(91)90078-5

Azov, Y., Goldman, J.C., 1982. Free ammonia inhibition of algal photosynthesis in intensive cultures. Appl. Environ. Microbiol. 43, 735–739.

Barbosa, M.J., Janssen, M., Ham, N., Tramper, J., Wijffels, R.H., 2003. Microalgae cultivation in air-lift reactors: Modeling biomass yield and growth rate as a function of mixing frequency. Biotechnol. Bioeng. 82, 170–179. https://doi.org/10.1002/bit.10563

Becker, E.W., 1994. Microalgae: Biotechnology and Microbiology. Cambridge University Press.

Bitton, G., 2005. Wastewater Microbiology, Third Edition.

Blanken, W., Postma, P.R., de Winter, L., Wijffels, R.H., Janssen, M., 2016. Predicting microalgae growth. Algal Res. 14, 28–38. https://doi.org/10.1016/j.algal.2015.12.020

Bonente, G., Pippa, S., Castellano, S., Bassi, R., Ballottari, M., 2012. Acclimation of *Chlamydomonas reinhardtii* to different growth irradiances. J. Biol. Chem. 287, 5833–5847. https://doi.org/10.1074/jbc.M111.304279

Borowitzka, M.A., 1998. Limits to Growth, in: Wong, Y.-S., Tam, N.F.Y. (Eds.), Wastewater treatment with algae, Biotechnology Intelligence Unit. Springer Berlin Heidelberg, pp. 203–226.

Brenner, K., You, L., Arnold, F.H., 2008. Engineering microbial consortia: a new frontier in synthetic biology. Trends Biotechnol. 26, 483–489. https://doi.org/10.1016/j.tibtech.2008.05.004

Buhr, H.O., Miller, S.B., 1983. A dynamic model of the high-rate algal-bacterial wastewater treatment pond. Water Res. 17, 29–37. https://doi.org/10.1016/0043-1354(83)90283-X

Cabanelas, I.T.D., Ruiz, J., Arbib, Z., Chinalia, F.A., Garrido-Pérez, C., Rogalla, F., Nascimento, I.A., Perales, J.A., 2013. Comparing the use of different domestic wastewaters for coupling microalgal production and nutrient removal. Bioresour. Technol. 131, 429–436. https://doi.org/10.1016/j.biortech.2012.12.152

Cai, T., Park, S.Y., Li, Y., 2013. Nutrient recovery from wastewater streams by microalgae: Status and prospects. Renew. Sustain. Energy Rev. 19, 360–369. https://doi.org/10.1016/j.rser.2012.11.030

Carvalho, A.P., Meireles, L.A., Malcata, F.X., 2006. Microalgal Reactors: A Review of Enclosed System Designs and Performances. Biotechnol. Prog. 22, 1490–1506. https://doi.org/10.1021/bp060065r

Chen, C.-Y., Yeh, K.-L., Aisyah, R., Lee, D.-J., Chang, J.-S., 2011. Cultivation, photobioreactor design and harvesting of microalgae for biodiesel production: A critical review. Bioresour. Technol. 102, 71–81. https://doi.org/10.1016/j.biortech.2010.06.159

Choi, O., Das, A., Yu, C.-P., Hu, Z., 2010. Nitrifying bacterial growth inhibition in the presence of algae and cyanobacteria. Biotechnol. Bioeng. 107, 1004–1011. https://doi.org/10.1002/bit.22860

Christenson, L., Sims, R., 2011. Production and harvesting of microalgae for wastewater treatment, biofuels, and bioproducts. Biotechnol. Adv. 29, 686–702. https://doi.org/10.1016/j.biotechadv.2011.05.015

Costache, T.A., Fernández, F.G.A., Morales, M.M., Fernández-Sevilla, J.M., Stamatin, I., Molina, E., 2013. Comprehensive model of microalgae photosynthesis rate as a function of culture conditions in photobioreactors. Appl. Microbiol. Biotechnol. 97, 7627–7637. https://doi.org/10.1007/s00253-013-5035-2

Cromar, N.J., Fallowfield, H.J., 1997. Effect of nutrient loading and retention time on performance of high rate algal ponds. J. Appl. Phycol. 9, 301–309. https://doi.org/10.1023/A:1007917610508

de-Bashan, L.E., Bashan, Y., 2004. Recent advances in removing phosphorus from wastewater and its future use as fertilizer (1997–2003). Water Res. 38, 4222–4246. https://doi.org/10.1016/j.watres.2004.07.014

de-Bashan, L.E., Hernandez, J.-P., Morey, T., Bashan, Y., 2004. Microalgae growth-promoting bacteria as "helpers" for microalgae: a novel approach for removing ammonium and phosphorus from municipal wastewater. Water Res. 38, 466–474. https://doi.org/10.1016/j.watres.2003.09.022

de Godos, I., Arbib, Z., Lara, E., Rogalla, F., 2016. Evaluation of high rate algae ponds for treatment of anaerobically digested wastewater: Effect of CO_2

addition and modification of dilution rate. Bioresour. Technol. 220, 253–261. https://doi.org/10.1016/j.biortech.2016.08.056

de Godos, I., Vargas, V.A., Guzmán, H.O., Soto, R., García, B., García, P.A., Muñoz, R., 2014. Assessing carbon and nitrogen removal in a novel anoxic–aerobic cyanobacterial–bacterial photobioreactor configuration with enhanced biomass sedimentation. Water Res. 61, 77–85. https://doi.org/10.1016/j.watres.2014.04.050

de la Noue, J., de Pauw, N., 1988. The potential of microalgal biotechnology: A review of production and uses of microalgae. Biotechnol. Adv. 6, 725–770. https://doi.org/10.1016/0734-9750(88)91921-0

de Mooij, T., de Vries, G., Latsos, C., Wijffels, R.H., Janssen, M., 2016. Impact of light color on photobioreactor productivity. Algal Res. 15, 32–42. https://doi.org/10.1016/j.algal.2016.01.015

Dębowski, M., Szwaja, S., Zieliński, M., Kisielewska, M., Stańczyk-Mazanek, E., 2017. The Influence of Anaerobic Digestion Effluents (ADEs) used as the nutrient sources for *Chlorella sp.* cultivation on fermentative biogas production. Waste Biomass Valorization 8, 1153–1161. https://doi.org/10.1007/s12649-016-9667-1

Decostere, B., De Craene, J., Van Hoey, S., Vervaeren, H., Nopens, I., Van Hulle, S.W.H., 2016. Validation of a microalgal growth model accounting with inorganic carbon and nutrient kinetics for wastewater treatment. Chem. Eng. J. 285, 189–197. https://doi.org/10.1016/j.cej.2015.09.111

Decostere, B., Janssens, N., Alvarado, A., Maere, T., Goethals, P., Van Hulle, S.W.H., Nopens, I., 2013. A combined respirometer–titrimeter for the determination of microalgae kinetics: Experimental data collection and modelling. Chem. Eng. J. 222, 85–93. https://doi.org/10.1016/j.cej.2013.01.103

Di Termini, I., Prassone, A., Cattaneo, C., Rovatti, M., 2011. On the nitrogen and phosphorus removal in algal photobioreactors. Ecol. Eng. 37, 976–980. https://doi.org/10.1016/j.ecoleng.2011.01.006

Dortch, Q., Clayton, J.R., Thoresen, S.S., Ahmed, S.I., 1984. Species differences in accumulation of nitrogen pools in phytoplankton. Mar. Biol. 81, 237–250. https://doi.org/10.1007/BF00393218

Droop, M.R., 1983. 25 Years of Algal Growth Kinetics A Personal View. Bot. Mar. 26, 99–112. https://doi.org/10.1515/botm.1983.26.3.99

Droop, M.R., 1973. Some thoughts on nutrient limitation in algae. J. Phycol. 9, 264–272. https://doi.org/10.1111/j.1529-8817.1973.tb04092.x

Edmundson, S.J., Huesemann, M.H., 2015. The dark side of algae cultivation: Characterizing night biomass loss in three photosynthetic algae, *Chlorella sorokiniana, Nannochloropsis salina and Picochlorum sp.* Algal Res. 12, 470–476. https://doi.org/10.1016/j.algal.2015.10.012

Ekama, G.A., Wentzel, M.C., 2008a. Nitrogen removal, in: biological wastewater treatment principles, modelling and design. IWA Publishing, pp. 245–272.

Ekama, G.A., Wentzel, M.C., 2008b. Organic material removal, in: biological wastewater treatment principles, modelling and design. IWA Publishing, pp. 245–272.

Escudero, A., Blanco, F., Lacalle, A., Pinto, M., 2014. Ammonium removal from anaerobically treated effluent by *Chlamydomonas acidophila*. Bioresour. Technol. 153, 62–68. https://doi.org/10.1016/j.biortech.2013.11.076

Evans, R.A., Cromar, N.J., Fallowfield, H.J., 2005. Performance of a pilot-scale high rate algal pond system treating abattoir wastewater in rural South Australia: nitrification and denitrification. Water Sci. Technol. 51, 117–124.

Falkowski, P.G., Raven, J.A., 2013. Aquatic Photosynthesis: Second Edition. Princeton University Press.

Fernández, F.G.A., Sevilla, J.M.F., Grima, E.M., 2013. Photobioreactors for the production of microalgae. Rev. Environ. Sci. Biotechnol. 12, 131–151. https://doi.org/10.1007/s11157-012-9307-6

Flynn, K.J., 1990. The determination of nitrogen status in microalgae. Mar. Ecol. Prog. Ser. 297–307.

Flynn, K.J., Davidson, K., Leftley, J.W., 1993. Carbon–nitrogen relations during batch growth of *Nannochloropsis oculata (eustigmatophyceae)* under alternating light and dark. J. Appl. Phycol. 5.

Fong, P., Foin, T.C., Zedler, J.B., 1994. A simulation model of lagoon algae based on nitrogen competition and internal storage. Ecol. Monogr. 64, 225–247. https://doi.org/10.2307/2937042

Fukami, K., Nishijima, T., Ishida, Y., 1997. Stimulative and inhibitory effects of bacteria on the growth of microalgae. Hydrobiologia 358, 185–191. https://doi.org/10.1023/A:1003139402315

García, D., Alcántara, C., Blanco, S., Pérez, R., Bolado, S., Muñoz, R., 2017. Enhanced carbon, nitrogen and phosphorus removal from domestic wastewater in a novel anoxic-aerobic photobioreactor coupled with biogas upgrading. Chem. Eng. J. 313, 424–434. https://doi.org/10.1016/j.cej.2016.12.054

García, J., Green, B.F., Lundquist, T., Mujeriego, R., Hernández-Mariné, M., Oswald, W.J., 2006. Long term diurnal variations in contaminant removal in high rate ponds treating urban wastewater. Bioresour. Technol. 97, 1709–1715. https://doi.org/10.1016/j.biortech.2005.07.019

García, J., Mujeriego, R., Hernández-Mariné, M., 2000. High rate algal pond operating strategies for urban wastewater nitrogen removal. J. Appl. Phycol. 12, 331–339. https://doi.org/10.1023/A:1008146421368

Gehring, T., Silva, J.D., Kehl, O., Castilhos, A.B., Jr, Costa, R.H.R., Uhlenhut, F., Alex, J., Horn, H., Wichern, M., 2010. Modelling waste stabilisation ponds with an extended version of ASM3. Water Sci. Technol. 61, 713–720. https://doi.org/10.2166/wst.2010.954

Gernaey, K.V., van Loosdrecht, M.C.., Henze, M., Lind, M., Jørgensen, S.B., 2004. Activated sludge wastewater treatment plant modelling and simulation: state of the art. Environ. Model. Softw. 19, 763–783. https://doi.org/10.1016/j.envsoft.2003.03.005

Godos, I. de, Blanco, S., García-Encina, P.A., Becares, E., Muñoz, R., 2009. Long-term operation of high rate algal ponds for the bioremediation of piggery wastewaters at high loading rates. Bioresour. Technol. 100, 4332–4339. https://doi.org/10.1016/j.biortech.2009.04.016

Godos, I. de, Vargas, V.A., Blanco, S., González, M.C.G., Soto, R., García-Encina, P.A., Becares, E., Muñoz, R., 2010. A comparative evaluation of microalgae for the degradation of piggery wastewater under photosynthetic oxygenation. Bioresour. Technol. 101, 5150–5158. https://doi.org/10.1016/j.biortech.2010.02.010

González, C., Marciniak, J., Villaverde, S., León, C., García, P.A., Muñoz, R., 2008. Efficient nutrient removal from swine manure in a tubular biofilm photo-bioreactor using algae-bacteria consortia. Water Sci. Technol. 58, 95. https://doi.org/10.2166/wst.2008.655

González-Fernández, C., Molinuevo-Salces, B., García-González, M.C., 2011a. Nitrogen transformations under different conditions in open ponds by means of microalgae–bacteria consortium treating pig slurry. Bioresour. Technol. 102, 960–966. https://doi.org/10.1016/j.biortech.2010.09.052

González-Fernández, C., Riaño-Irazábal, B., Molinuevo-Salces, B., Blanco, S., García-González, M., 2011b. Effect of operational conditions on the degradation of organic matter and development of microalgae–bacteria consortia when treating swine slurry. Appl. Microbiol. Biotechnol. 90, 1147–1153. https://doi.org/10.1007/s00253-011-3111-z

Gonzalez-Martinez, A., Rodriguez-Sanchez, A., Lotti, T., Garcia-Ruiz, M.-J., Osorio, F., Gonzalez-Lopez, J., Loosdrecht, M.C.M. van, 2016. Comparison of bacterial communities of conventional and A-stage activated sludge systems. Sci. Rep. 6, srep18786. https://doi.org/10.1038/srep18786

Granados, M.R., Acién, F.G., Gómez, C., Fernández-Sevilla, J.M., Molina Grima, E., 2012. Evaluation of flocculants for the recovery of freshwater microalgae.

Bioresour. Technol. 118, 102–110. https://doi.org/10.1016/j.biortech.2012.05.018

Grobbelaar, J.U., 2008. Algal Nutrition - Mineral Nutrition, in: Handbook of Microalgal Culture: Biotechnology and Applied Phycology. John Wiley & Sons.

Gujer, W., Henze, M., Mino, T., Loosdrecht, M. van, 1999. Activated sludge model No. 3. Water Sci. Technol. 39, 183–193. https://doi.org/10.1016/S0273-1223(98)00785-9

Gutzeit, G., Lorch, D., Weber, A., Engels, M., Neis, U., 2005. Bioflocculent algal-bacterial biomass improves low-cost wastewater treatment. Water Sci. Technol. 52, 9–18.

Halfhide, T., Dalrymple, O.K., Wilkie, A.C., Trimmer, J., Gillie, B., Udom, I., Zhang, Q., Ergas, S.J., 2015. Growth of an indigenous algal consortium on anaerobically digested municipal sludge centrate: photobioreactor performance and modeling. BioEnergy Res. 8, 249–258. https://doi.org/10.1007/s12155-014-9513-x

He, P.J., Mao, B., Shen, C.M., Shao, L.M., Lee, D.J., Chang, J.S., 2013. Cultivation of *Chlorella vulgaris* on wastewater containing high levels of ammonia for biodiesel production. Bioresour. Technol. 129, 177–181. https://doi.org/10.1016/j.biortech.2012.10.162

Hellebust, J.A., Ahmad, I., 1989. Regulation of nitrogen assimilation in green microalgae. Biol. Oceanogr. 6, 241–255. https://doi.org/10.1080/01965581.1988.10749529

Henderson, R., Parsons, S.A., Jefferson, B., 2008. The impact of algal properties and pre-oxidation on solid–liquid separation of algae. Water Res. 42, 1827–1845. https://doi.org/10.1016/j.watres.2007.11.039

Henze, M., 2000. Activated Sludge Models ASM1, ASM2, ASM2d and ASM3. IWA Publishing.

Hernández, D., Riaño, B., Coca, M., García-González, M.C., 2013. Treatment of agro-industrial wastewater using microalgae–bacteria consortium combined with anaerobic digestion of the produced biomass. Bioresour. Technol., Biorefineries 135, 598–603. https://doi.org/10.1016/j.biortech.2012.09.029

Hoffmann, J.P., 1998. Wastewater treatment with suspended and nonsuspended Algae. J. Phycol. 34, 757–763. https://doi.org/10.1046/j.1529-8817.1998.340757.x

Holenda, B., Domokos, E., Rédey, Á., Fazakas, J., 2008. Dissolved oxygen control of the activated sludge wastewater treatment process using model predictive control. Comput. Chem. Eng. 32, 1270–1278. https://doi.org/10.1016/j.compchemeng.2007.06.008

Iacopozzi, I., Innocenti, V., Marsili-Libelli, S., Giusti, E., 2007. A modified activated sludge model No. 3 (ASM3) with two-step nitrification-denitrification. Env. Model Softw 22, 847–861. https://doi.org/10.1016/j.envsoft.2006.05.009

Janssen, M., Lamers, P., 2013. Microalgae Biotechnology Notes.

Jeon, Y.-C., Cho, C.-W., Yun, Y.-S., 2005. Measurement of microalgal photosynthetic activity depending on light intensity and quality. Biochem. Eng. J. 27, 127–131. https://doi.org/10.1016/j.bej.2005.08.017

Kaelin, D., Manser, R., Rieger, L., Eugster, J., Rottermann, K., Siegrist, H., 2009. Extension of ASM3 for two-step nitrification and denitrification and its calibration and validation with batch tests and pilot scale data. Water Res. 43, 1680–1692. https://doi.org/10.1016/j.watres.2008.12.039

Kapdan, I.K., Aslan, S., 2008. Application of the Stover–Kincannon kinetic model to nitrogen removal by *Chlorella vulgaris* in a continuously operated immobilized photobioreactor system. J. Chem. Technol. Biotechnol. 83, 998–1005. https://doi.org/10.1002/jctb.1905

Karya, N.G.A.I., van der Steen, N.P., Lens, P.N.L., 2013. Photo-oxygenation to support nitrification in an algal-bacterial consortium treating artificial wastewater. Bioresour. Technol. 134, 244–250. https://doi.org/10.1016/j.biortech.2013.02.005

Kayombo, S., Mbwette, T.S.., Mayo, A.., Katima, J.H.., Jorgensen, S.., 2000. Modelling diurnal variation of dissolved oxygen in waste stabilization ponds. Ecol. Model. 127, 21–31. https://doi.org/10.1016/S0304-3800(99)00196-9

Khan, A.A., Gaur, R.Z., Tyagi, V.K., Khursheed, A., Lew, B., Mehrotra, I., Kazmi, A.A., 2011. Sustainable options of post treatment of UASB effluent treating sewage: A review. Resour. Conserv. Recycl. 55, 1232–1251. https://doi.org/10.1016/j.resconrec.2011.05.017

Kim, T.-H., Lee, Y., Han, S.-H., Hwang, S.-J., 2013. The effects of wavelength and wavelength mixing ratios on microalgae growth and nitrogen, phosphorus removal using *Scenedesmus sp.* for wastewater treatment. Bioresour. Technol. 130, 75–80. https://doi.org/10.1016/j.biortech.2012.11.134

Kinyua, M.N., Cunningham, J., Ergas, S.J., 2014. Effect of solids retention time on the bioavailability of organic carbon in anaerobically digested swine waste. Bioresour. Technol. 162, 14–20. https://doi.org/10.1016/j.biortech.2014.03.111

Kirkwood, A.E., Nalewajko, C., Fulthorpe, R.R., 2006. The effects of cyanobacterial exudates on bacterial growth and biodegradation of organic contaminants. Microb. Ecol. 51, 4–12. https://doi.org/10.1007/s00248-004-0058-y

Kouba, V., Catrysse, M., Stryjova, H., Jonatova, I., Volcke, E.I.P., Svehla, P., Bartacek, J., 2014. The impact of influent total ammonium nitrogen concentration on nitrite-oxidizing bacteria inhibition in moving bed biofilm reactor. Water Sci. Technol. J. Int. Assoc. Water Pollut. Res. 69, 1227–1233. https://doi.org/10.2166/wst.2013.757

Kouzuma, A., Watanabe, K., 2015. Exploring the potential of algae/bacteria interactions. Curr. Opin. Biotechnol. 33, 125–129. https://doi.org/10.1016/j.copbio.2015.02.007

Larsdotter, K., 2006. Wastewater treatnment with microalgae - a literature review. Vatten 31–38.

Lavín, P.L., Lourenço, S.O., 2005. An evaluation of the accumulation of intracellular inorganic nitrogen pools by marine microalgae in batch cultures. Braz. J. Oceanogr. 53, 55–68. https://doi.org/10.1590/S1679-87592005000100006

Lee, J., Cho, D.-H., Ramanan, R., Kim, B.-H., Oh, H.-M., Kim, H.-S., 2013. Microalgae-associated bacteria play a key role in the flocculation of *Chlorella vulgaris*. Bioresour. Technol. 131, 195–201. https://doi.org/10.1016/j.biortech.2012.11.130

Liang, Z., Liu, Y., Ge, F., Xu, Y., Tao, N., Peng, F., Wong, M., 2013. Efficiency assessment and pH effect in removing nitrogen and phosphorus by algae-bacteria combined system of *Chlorella vulgaris* and *Bacillus licheniformis*. Chemosphere 92, 1383–1389. https://doi.org/10.1016/j.chemosphere.2013.05.014

Liu, G., Wang, J., 2012. Probing the stoichiometry of the nitrification process using the respirometric approach. Water Res. 46, 5954–5962. https://doi.org/10.1016/j.watres.2012.08.006

Liu, J., Wu, Y., Wu, C., Muylaert, K., Vyverman, W., Yu, H.-Q., Muñoz, R., Rittmann, B., 2017. Advanced nutrient removal from surface water by a consortium of attached microalgae and bacteria: A review. Bioresour. Technol. 241, 1127–1137. https://doi.org/10.1016/j.biortech.2017.06.054

Manser, N.D., Wang, M., Ergas, S.J., Mihelcic, J.R., Mulder, A., van de Vossenberg, J., van Lier, J.B., van der Steen, P., 2016. Biological nitrogen removal in a photosequencing batch reactor with an algal-nitrifying bacterial consortium and anammox granules. Environ. Sci. Technol. Lett. 3, 175–179. https://doi.org/10.1021/acs.estlett.6b00034

Manser, R., Gujer, W., Siegrist, H., 2005. Consequences of mass transfer effects on the kinetics of nitrifiers. Water Res. 39, 4633–4642. https://doi.org/10.1016/j.watres.2005.09.020

Mara, D., 2004. Domestic Wastewater treatment in developing countries. Earthscan, London.

Martínez, M.E., Jiménez, J.M., El Yousfi, F., 1999. Influence of phosphorus concentration and temperature on growth and phosphorus uptake by the microalga *Scenedesmus obliquus*. Bioresour. Technol. 67, 233–240. https://doi.org/10.1016/S0960-8524(98)00120-5

Martinez Sancho, M.E., Bravo Rodriguez, V., Sanchez Villasclaras, S., Molina Grima, E., 1991. Determining the kinetic parameters characteristic of microalgal Growth. Chem. Eng. Educ. 25, 145–49.

Maza-Márquez, P., González-Martínez, A., Rodelas, B., González-López, J., 2017. Full-scale photobioreactor for biotreatment of olive washing water: Structure and diversity of the microalgae-bacteria consortium. Bioresour. Technol. 238, 389–398. https://doi.org/10.1016/j.biortech.2017.04.048

Medina, M., Neis, U., 2007. Symbiotic algal bacterial wastewater treatment: effect of food to microorganism ratio and hydraulic retention time on the process performance. Water Sci. Technol. 55, 165. https://doi.org/10.2166/wst.2007.351

Metcalf & Eddy, Tchobanoglous, G., Burton, F.L., Stensel, H.D., 2002. Wastewater engineering: treatment and reuse, Edición: 4. ed. McGraw-Hill Higher Education, Boston, Mass.

Mokashi, K., Shetty, V., George, S.A., Sibi, G., 2016. Sodium Bicarbonate as inorganic carbon source for higher biomass and lipid production integrated carbon capture in *Chlorella vulgaris*. Achiev. Life Sci. 10, 111–117. https://doi.org/10.1016/j.als.2016.05.011

Molina Grima, E., Camacho, F.G., Pérez, J.A.S., Sevilla, J.M.F., Fernández, F.G.A., Gómez, A.C., 1994. A mathematical model of microalgal growth in light-limited chemostat culture. J. Chem. Technol. Biotechnol. 61, 167–173. https://doi.org/10.1002/jctb.280610212

Molinuevo-Salces, B., García-González, M.C., González-Fernández, C., 2010. Performance comparison of two photobioreactors configurations (open and closed to the atmosphere) treating anaerobically degraded swine slurry. Bioresour. Technol. 101, 5144–5149. https://doi.org/10.1016/j.biortech.2010.02.006

Mooij, P.R., Graaff, D.R. de, Loosdrecht, M.C.M. van, Kleerebezem, R., 2015. Starch productivity in cyclically operated photobioreactors with marine microalgae—effect of ammonium addition regime and volume exchange ratio. J. Appl. Phycol. 27, 1121–1126. https://doi.org/10.1007/s10811-014-0430-3

Mukarunyana, B., van de Vossenberg, J., van Lier, J.B., van der Steen, P., 2018. Photo-oxygenation for nitritation and the effect of dissolved oxygen concentrations on anaerobic ammonium oxidation. Sci. Total Environ. 634, 868–874. https://doi.org/10.1016/j.scitotenv.2018.04.082

Muñoz, R., Guieysse, B., 2006. Algal–bacterial processes for the treatment of hazardous contaminants: A review. Water Res. 40, 2799–2815. https://doi.org/10.1016/j.watres.2006.06.011

Munz, G., Lubello, C., Oleszkiewicz, J.A., 2011. Factors affecting the growth rates of ammonium and nitrite oxidizing bacteria. Chemosphere 83, 720–725. https://doi.org/10.1016/j.chemosphere.2011.01.058

Noyola, A., Padilla-Rivera, A., Morgan-Sagastume, J.M., Güereca, L.P., Hernández-Padilla, F., 2012. Typology of municipal wastewater treatment technologies in latin america. CLEAN – Soil Air Water 40, 926–932. https://doi.org/10.1002/clen.201100707

Olguín, E.J., 2003. Phycoremediation: key issues for cost-effective nutrient removal processes. Biotechnol. Adv. 22, 81–91. https://doi.org/10.1016/S0734-9750(03)00130-7

Osada, T., Haga, K., Harada, Y., 1991. Removal of nitrogen and phosphorus from swine wastewater by the activated sludge units with the intermittent aeration process. Water Res. 25, 1377–1388. https://doi.org/10.1016/0043-1354(91)90116-8

Oswald, W.J., Gotaas, H.B., Ludwig, H.F., Lynch, V., 1953. Algae Symbiosis in Oxidation Ponds: III. Photosynthetic Oxygenation. Sew. Ind. Wastes Vol. 25, 692–705.

Park, J., Jin, H.-F., Lim, B.-R., Park, K.-Y., Lee, K., 2010. Ammonia removal from anaerobic digestion effluent of livestock waste using green alga *Scenedesmus sp.* Bioresour. Technol. 101, 8649–8657. https://doi.org/10.1016/j.biortech.2010.06.142

Park, J.B.K., Craggs, R.J., 2011. Nutrient removal in wastewater treatment high rate algal ponds with carbon dioxide addition. Water Sci. Technol. 63, 1758. https://doi.org/10.2166/wst.2011.114

Park, J.B.K., Craggs, R.J., 2010. Wastewater treatment and algal production in high rate algal ponds with carbon dioxide addition. Water Sci. Technol. 61, 633. https://doi.org/10.2166/wst.2010.951

Perez-Garcia, O., Escalante, F.M.E., de-Bashan, L.E., Bashan, Y., 2011. Heterotrophic cultures of microalgae: Metabolism and potential products. Water Res. 45, 11–36. https://doi.org/10.1016/j.watres.2010.08.037

Plappally, A.K., Lienhard, J.H., 2012. Energy requirements for water production, treatment, end use, reclamation, and disposal. Renew. Sustain. Energy Rev. 16, 4818–4848. https://doi.org/10.1016/j.rser.2012.05.022

Pollice, A., Tandoi, V., Lestingi, C., 2002. Influence of aeration and sludge retention time on ammonium oxidation to nitrite and nitrate. Water Res. 36, 2541–2546. https://doi.org/10.1016/S0043-1354(01)00468-7

Posadas, E., García-Encina, P.-A., Soltau, A., Domínguez, A., Díaz, I., Muñoz, R., 2013. Carbon and nutrient removal from centrates and domestic wastewater using algal–bacterial biofilm bioreactors. Bioresour. Technol. 139, 50–58. https://doi.org/10.1016/j.biortech.2013.04.008

Powell, N., Shilton, A.N., Pratt, S., Chisti, Y., 2008. Factors influencing luxury uptake of phosphorus by microalgae in waste stabilization ponds. Environ. Sci. Technol. 42, 5958–5962. https://doi.org/10.1021/es703118s

Quijano, G., Arcila, J.S., Buitrón, G., 2017. Microalgal-bacterial aggregates: Applications and perspectives for wastewater treatment. Biotechnol. Adv. 35, 772–781. https://doi.org/10.1016/j.biotechadv.2017.07.003

Quinn, J., de Winter, L., Bradley, T., 2011. Microalgae bulk growth model with application to industrial scale systems. Bioresour. Technol. 102, 5083–5092. https://doi.org/10.1016/j.biortech.2011.01.019

Rada-Ariza, A.., Rahman, A., Zalivina, N., Lopez-Vazquez, C.., Van der Steen, N.P., Lens, P.N.L., 2015. Sludge retention time effects on ammonium removal in a photo-CSTR using a microalgae-bacteria consortium. Presented at the IWA Specialist Conference on Nutrient Removal and Recovery: moving innovation into practice, IWA Pub., Gdansk, Poland.

Rada-Ariza, María, A., Lopez-Vazquez, C.M., Van der Steen, N.P., Lens, P.N.L., 2017. Nitrification by microalgal-bacterial consortia for ammonium removal in flat panel sequencing batch photo-bioreactors. Bioresour. Technol. 245, 81 - 89. https://doi.org/10.1016/j.biortech.2017.08.019

Ramanan, R., Kim, B.-H., Cho, D.-H., Oh, H.-M., Kim, H.-S., 2016. Algae–bacteria interactions: Evolution, ecology and emerging applications. Biotechnol. Adv. 34, 14–29. https://doi.org/10.1016/j.biotechadv.2015.12.003

Rawat, I., Ranjith Kumar, R., Mutanda, T., Bux, F., 2011. Dual role of microalgae: Phycoremediation of domestic wastewater and biomass production for sustainable biofuels production. Appl. Energy 88, 3411–3424. https://doi.org/10.1016/j.apenergy.2010.11.025

Reichert, P., 1998. AQUASIM 2.0 - Tutorial. Computer program for the identification and simulation of aquatic systems.

Reichert, P., 1994. Aquasim – a tool for simulation and data analysis of aquatic systems. Water Sci. Technol. 30, 21–30.

Reichert, P., Borchardt, D., Henze, M., Rauch, W., Shanahan, P., Somlyódy, L., Vanrolleghem, P., 2001. River Water Quality Model no. 1 (RWQM1): II. Biochemical process equations. Water Sci. Technol. 43, 11–30.

Riaño, B., Hernández, D., García-González, M.C., 2012. Microalgal-based systems for wastewater treatment: Effect of applied organic and nutrient loading rate on biomass composition. Ecol. Eng. 49, 112–117. https://doi.org/10.1016/j.ecoleng.2012.08.021

Risgaard-Petersen, N., Nicolaisen, M.H., Revsbech, N.P., Lomstein, B.A., 2004. Competition between Ammonia-Oxidizing bacteria and benthic microalgae. Appl. Environ. Microbiol. 70, 5528–5537. https://doi.org/10.1128/AEM.70.9.5528-5537.2004

Ruiz, J., Álvarez, P., Arbib, Z., Garrido, C., Barragán, J., Perales, J.A., 2011. Effect of nitrogen and phosphorus concentration on their removal kinetic in treated urban wastewater by *Chlorella Vulgaris*. Int. J. Phytoremediation 13, 884–896. https://doi.org/10.1080/15226514.2011.573823

Ruiz-Martinez, A., Martin Garcia, N., Romero, I., Seco, A., Ferrer, J., 2012. Microalgae cultivation in wastewater: nutrient removal from anaerobic membrane bioreactor effluent. Bioresour. Technol. 126, 247–253. https://doi.org/10.1016/j.biortech.2012.09.022

Safonova, E., Kvitko, K. v., Iankevitch, M. i., Surgko, L. f., Afti, I. a., Reisser, W., 2004. Biotreatment of industrial wastewater by selected algal-bacterial consortia. Eng. Life Sci. 4, 347–353. https://doi.org/10.1002/elsc.200420039

Samorì, G., Samorì, C., Guerrini, F., Pistocchi, R., 2013. Growth and nitrogen removal capacity of Desmodesmus communis and of a natural microalgae consortium in a batch culture system in view of urban wastewater treatment: part I. Water Res. 47, 791–801. https://doi.org/10.1016/j.watres.2012.11.006

Shilton, A., 2006. Pond Treatment Technology. IWA Publishing, London, United Kingdom.

Sin, G., Guisasola, A., Pauw, D.J.W.D., Baeza, J.A., Carrera, J., Vanrolleghem, P.A., 2005. A new approach for modelling simultaneous storage and growth processes for activated sludge systems under aerobic conditions. Biotechnol. Bioeng. 92, 600–613. https://doi.org/10.1002/bit.20741

Sivakumar, G., Xu, J., Thompson, R.W., Yang, Y., Randol-Smith, P., Weathers, P.J., 2012. Integrated green algal technology for bioremediation and biofuel. Bioresour. Technol. 107, 1–9. https://doi.org/10.1016/j.biortech.2011.12.091

Solimeno, A., Parker, L., Lundquist, T., García, J., 2017. Integral microalgae-bacteria model (BIO_ALGAE): Application to wastewater high rate algal ponds. Sci. Total Environ. 601–602, 646–657. https://doi.org/10.1016/j.scitotenv.2017.05.215

Solimeno, A., Samsó, R., Uggetti, E., Sialve, B., Steyer, J.-P., Gabarró, A., García, J., 2015. New mechanistic model to simulate microalgae growth. Algal Res. 12, 350–358. https://doi.org/10.1016/j.algal.2015.09.008

Spanjers, H., Vanrolleghem, P.A., 2016. Respirometry, in: experimental Methods in wastewater treatment. IWA Pub, London, pp. 133–176.

Su, Y., Mennerich, A., Urban, B., 2012a. Synergistic cooperation between wastewater-born algae and activated sludge for wastewater treatment: Influence of algae and sludge inoculation ratios. Bioresour. Technol. 105, 67–73. https://doi.org/10.1016/j.biortech.2011.11.113

Su, Y., Mennerich, A., Urban, B., 2012b. Comparison of nutrient removal capacity and biomass settleability of four high-potential microalgal species. Bioresour. Technol. 124, 157–162. https://doi.org/10.1016/j.biortech.2012.08.037

Su, Y., Mennerich, A., Urban, B., 2011. Municipal wastewater treatment and biomass accumulation with a wastewater-born and settleable algal-bacterial culture. Water Res. 45, 3351–3358. https://doi.org/10.1016/j.watres.2011.03.046

Subashchandrabose, S.R., Ramakrishnan, B., Megharaj, M., Venkateswarlu, K., Naidu, R., 2011. Consortia of cyanobacteria/microalgae and bacteria: Biotechnological potential. Biotechnol. Adv. 29, 896–907. https://doi.org/10.1016/j.biotechadv.2011.07.009

Sutherland, D.L., Montemezzani, V., Howard-Williams, C., Turnbull, M.II., Broady, P.A., Craggs, R.J., 2015. Modifying the high rate algal pond light environment and its effects on light absorption and photosynthesis. Water Res. 70, 86–96. https://doi.org/10.1016/j.watres.2014.11.050

Sutherland, D.L., Turnbull, M.H., Broady, P.A., Craggs, R.J., 2014. Effects of two different nutrient loads on microalgal production, nutrient removal and photosynthetic efficiency in pilot-scale wastewater high rate algal ponds. Water Res. 66, 53–62. https://doi.org/10.1016/j.watres.2014.08.010

Takabe, Y., Hidaka, T., Tsumori, J., Minamiyama, M., 2016. Effects of hydraulic retention time on cultivation of indigenous microalgae as a renewable energy source using secondary effluent. Bioresour. Technol. 207, 399–408. https://doi.org/10.1016/j.biortech.2016.01.132

Tan, X.-B., Zhao, X.-C., Yang, L.-B., Liao, J.-Y., Zhou, Y.-Y., 2018. Enhanced biomass and lipid production for cultivating *Chlorella pyrenoidosa* in anaerobically digested starch wastewater using various carbon sources and up-scaling culture outdoors. Biochem. Eng. J. 135, 105–114. https://doi.org/10.1016/j.bej.2018.04.005

Tandukar, M., Ohashi, A., Harada, H., 2007. Performance comparison of a pilot-scale UASB and DHS system and activated sludge process for the treatment of municipal wastewater. Water Res. 41, 2697–2705. https://doi.org/10.1016/j.watres.2007.02.027

Taylor, R.P., Jones, C.L.W., Laing, M., Dames, J., 2018. The potential use of treated brewery effluent as a water and nutrient source in irrigated crop production. Water Resour. Ind. 19, 47–60. https://doi.org/10.1016/j.wri.2018.02.001

Tiron, O., Bumbac, C., Manea, E., Stefanescu, M., Lazar, M.N., 2017. Overcoming microalgae harvesting barrier by activated algae granules. Sci. Rep. 7, 4646. https://doi.org/10.1038/s41598-017-05027-3

Tuantet, K., 2015. Microalgae cultivation for nutrient recovery from human urine. Wageningen University, Wageningen.

Tuantet, K., Janssen, M., Temmink, H., Zeeman, G., Wijffels, R.H., Buisman, C.J.N., 2013. Microalgae growth on concentrated human urine. J. Appl. Phycol. 26, 287–297. https://doi.org/10.1007/s10811-013-0108-2

Tuantet, K., Temmink, H., Zeeman, G., Janssen, M., Wijffels, R.H., Buisman, C.J.N., 2014. Nutrient removal and microalgal biomass production on urine in a short light-path photobioreactor. Water Res. 55, 162–174. https://doi.org/10.1016/j.watres.2014.02.027

UNESCO, 2017. Wastewater: The Untapped Resource.

Unnithan, V.V., Unc, A., Smith, G.B., 2014. Mini-review: A priori considerations for bacteria–algae interactions in algal biofuel systems receiving municipal wastewaters. Algal Res. 4, 35–40. https://doi.org/10.1016/j.algal.2013.11.009

Vadivelu, V.M., Keller, J., Yuan, Z., 2007. Free ammonia and free nitrous acid inhibition on the anabolic and catabolic processes of Nitrosomonas and Nitrobacter. Water Sci. Technol. 56, 89–97. https://doi.org/10.2166/wst.2007.612

Valigore, J.M., Gostomski, P.A., Wareham, D.G., O'Sullivan, A.D., 2012. Effects of hydraulic and solids retention times on productivity and settleability of microbial (microalgal-bacterial) biomass grown on primary treated wastewater as a biofuel feedstock. Water Res. 46, 2957–2964. https://doi.org/10.1016/j.watres.2012.03.023

Van Den Hende, S., 2014. Microalgal bacterial flocs for wastewater treatment: from concept to pilot scale (dissertation). Ghent University.

Van Den Hende, S., Carré, E., Cocaud, E., Beelen, V., Boon, N., Vervaeren, H., 2014. Treatment of industrial wastewaters by microalgal bacterial flocs in sequencing batch reactors. Bioresour. Technol. 161, 245–254. https://doi.org/10.1016/j.biortech.2014.03.057

Van Den Hende, S., Vervaeren, H., Saveyn, H., Maes, G., Boon, N., 2011. Microalgal bacterial floc properties are improved by a balanced inorganic/organic carbon ratio. Biotechnol. Bioeng. 108, 549–558. https://doi.org/10.1002/bit.22985

van der Steen, P., Rahsilawati, K., Rada-Ariza, A.M., Lopez-Vazquez, C.M., Lens, P.N.L., 2015. A new photo-activated sludge system for nitrification by an algal-bacterial consortium in a photo-bioreactor with biomass recycle. Water Sci. Technol. 72, 443–450. https://doi.org/10.2166/wst.2015.205

van Loosdrecht, M.C.M., Ekama, G.A., Wentzel, M.C., Brdjanovic, D., Hooijmans, C.M., 2008. Modelling activated sludge processes, in: biological wastewater treatment principles, modelling and design. IWA Publishing, pp. 245–272.

Vargas, G., Donoso-Bravo, A., Vergara, C., Ruiz-Filippi, G., 2016. Assessment of microalgae and nitrifiers activity in a consortium in a continuous operation and the effect of oxygen depletion. Electron. J. Biotechnol. 23, 63–68. https://doi.org/10.1016/j.ejbt.2016.08.002

von Sperling, M., Chernicharo, C., 2002. Urban wastewater treatment technologies and the implementation of discharge standards in developing countries. Urban Water 4, 105–114. https://doi.org/10.1016/S1462-0758(01)00066-8

Wágner, D.S., Valverde-Pérez, B., Sæbø, M., Bregua de la Sotilla, M., Van Wagenen, J., Smets, B.F., Plósz, B.G., 2016. Towards a consensus-based biokinetic model for green microalgae – The ASM-A. Water Res. 103, 485–499. https://doi.org/10.1016/j.watres.2016.07.026

Wang, B., Lan, C.Q., Horsman, M., 2012. Closed photobioreactors for production of microalgal biomasses. Biotechnol. Adv. 30, 904–912. https://doi.org/10.1016/j.biotechadv.2012.01.019

Wang, L., Li, Y., Chen, P., Min, M., Chen, Y., Zhu, J., Ruan, R.R., 2010. Anaerobic digested dairy manure as a nutrient supplement for cultivation of oil-rich green microalgae *Chlorella sp.* Bioresour. Technol. 101, 2623–2628. https://doi.org/10.1016/j.biortech.2009.10.062

Wang, M., Keeley, R., Zalivina, N., Halfhide, T., Scott, K., Zhang, Q., van der Steen, P., Ergas, S.J., 2018. Advances in algal-prokaryotic wastewater treatment: A review of nitrogen transformations, reactor configurations and molecular tools. J. Environ. Manage. 217, 845–857. https://doi.org/10.1016/j.jenvman.2018.04.021

Wang, M., Park, C., 2015. Investigation of anaerobic digestion of *Chlorella sp.* and *Micractinium sp.* grown in high-nitrogen wastewater and their co-digestion with waste activated sludge. Biomass Bioenergy 80, 30–37. https://doi.org/10.1016/j.biombioe.2015.04.028

Wang, M., Yang, H., Ergas, S.J., van der Steen, P., 2015. A novel shortcut nitrogen removal process using an algal-bacterial consortium in a photo-sequencing batch reactor (PSBR). Water Res. 87, 38–48. https://doi.org/10.1016/j.watres.2015.09.016

Wang, X., Hao, C., Zhang, F., Feng, C., Yang, Y., 2011. Inhibition of the growth of two blue-green algae species (*Microsystis aruginosa and Anabaena spiroides)* by acidification treatments using carbon dioxide. Bioresour. Technol. 102, 5742–5748. https://doi.org/10.1016/j.biortech.2011.03.015

Wiesmann, U., 1994a. Biological nitrogen removal from wastewater. Adv. Biochem. Eng. Biotechnol. 51, 113–154.

Wiesmann, U., 1994b. Biological nitrogen removal from wastewater, in: Biotechnics/Wastewater, Advances in Biochemical Engineering/Biotechnology. Springer Berlin Heidelberg, pp. 113–154.

Willmott, C.J., Robeson, S.M., Matsuura, K., 2012. A refined index of model performance. Int. J. Climatol. 32, 2088–2094. https://doi.org/10.1002/joc.2419

Wolf, G., Picioreanu, C., van Loosdrecht, M.C.M., 2007. Kinetic modeling of phototrophic biofilms: The PHOBIA model. Biotechnol. Bioeng. 97, 1064–1079. https://doi.org/10.1002/bit.21306

Wu, J., He, C., van Loosdrecht, M.C.M., Pérez, J., 2016. Selection of ammonium oxidizing bacteria (AOB) over nitrite oxidizing bacteria (NOB) based on conversion rates. Chem. Eng. J. 304, 953–961. https://doi.org/10.1016/j.cej.2016.07.019

Wu, Y.-H., Hu, H.-Y., Yu, Y., Zhang, T.-Y., Zhu, S.-F., Zhuang, L.-L., Zhang, X., Lu, Y., 2014. Microalgal species for sustainable biomass/lipid production using wastewater as resource: A review. Renew. Sustain. Energy Rev. 33, 675–688. https://doi.org/10.1016/j.rser.2014.02.026

Yoshioka, T., Saijo, Y., 1984. Photoinhibition and recovery of NH_4^+-oxidizing bacteria and NO_2^-–oxidizing bacteria. J Gen Appl Microbiol 30, 151–166. https://doi.org/10.2323/jgam.30.151

Young, D.F., Koopman, B., 1991. Electricity uste in small wastewater treatment plants. J. Environ. Eng. 117, 300–307. https://doi.org/10.1061/(ASCE)0733-9372(1991)117:3(300)

Yun, Y.-S., Park, J.M., 2003. Kinetic modeling of the light-dependent photosynthetic activity of the green microalga *Chlorella vulgaris*. Biotechnol. Bioeng. 83, 303–311. https://doi.org/10.1002/bit.10669

Zalivina, N., 2014. The effect of SRT and pH on nitrogen removal in the photo-activated sludge system. UNESCO-IHE, Delft.

Zambrano, J., Krustok, I., Nehrenheim, E., Carlsson, B., 2016. A simple model for algae-bacteria interaction in photo-bioreactors. Algal Res. 19, 155–161. https://doi.org/10.1016/j.algal.2016.07.022

Zhao, X., Zhou, Y., Huang, S., Qiu, D., Schideman, L., Chai, X., Zhao, Y., 2014. Characterization of microalgae-bacteria consortium cultured in landfill leachate for carbon fixation and lipid production. Bioresour. Technol. 156, 322–328. https://doi.org/10.1016/j.biortech.2013.12.112

Zhou, Q., He, S.L., He, X.J., Huang, X.F., Picot, B., Li, X.D., Chen, G., 2006. Nutrients removal mechanisms in high rate algal pond treating rural domestic sewage in East China. Water Sci. Technol. 6, 43. https://doi.org/10.2166/ws.2006.956

List of acronyms

AD	Anaerobic Digestion
ADS	Anaerobic Digested Slurry
ANOVA	Analysis of Variance
AOA	Ammonium Oxidising Archaea
AOB	Ammonium-Oxidising Bacteria
ARE	Ammonium Removal Efficiency
ARR	Ammonium Removal Rate
AS	Activated Sludge
ASM	Activated Sludge Model
ATU	N-Allylthiourea
BOD	Biological Oxygen Demand
COD	Chemical Oxygen Demand
DO	Dissolved Oxygen
EPS	Extra-Polymer Substances
FNA	Free Nitrous Acid
FPR	Flat-Panel Reactor
FSA	Free and Saline Ammonia
HRAP	High Rate Algae Ponds

HRT	Hydraulic Retention Time
IOA	Index of Agreement
IWA	International Water Association
LED	Light Emitting Diode
MBR	Membrane Bioreactor
MGPB	Microalgae Growth-Promoting Bacteria
NLR	Nitrogen Loading Rate
NOB	Nitrite Oxidising Bacteria
NRE	Nitrogen Removal Efficiency
OHO	Ordinary Heterotrophic Bacteria
PAR	Photosynthetically-Active Radiation
PAS	Photo-Activated Sludge system
PBS	Phosphate Buffer Solution
PSBR	Photo sequencing Batch Reactor
RMSE	Root-Mean-Square Error
RT	Respirometric Test
SAT	Soil Aquifer Treatment
SBR	Sequencing Batch Reactor
SRT	Solid Retention Time
SVI	Sludge Volume Index

TIN	Total Inorganic Nitrogen
TN	Total Nitrogen
TSS	Total Suspended Solids
UASB	Upflow Anaerobic Sludge Blanket
VSS	Volatile Suspended Solids
WSP	Waste Stabilization Ponds

List of Tables

Table 2.1 Nutrient removal using a microalgal-bacterial consortia for different types of wastewater and using different types of reactors. Source: Subashchandrabose et al. (2011). 22

Table 3.1. Operational conditions and nitrogen (ammonium) loading rate (NLR) in each experimental periods applied to the two flat panel reactors........ 44

Table 3.2. Solids retention time (days) in FPR1 and FPR2 during periods 3 to 7. 53

Table 4.1. TSS, VSS, Chl-a concentration, Chl-a content in the biomass, and biomass productivity in the 4 periods (*Values reported for the biomass inside the photobioreactor)........ 84

Table 4.2. Ammonium removal efficiencies and rates for the 4 operational periods 86

Table 4.3. Nitrogen removed by the different mechanisms for the different operational periods........ 95

Table 4.4. Total oxygen produced and consumed in the SBR........ 97

Table 4.5. Oxygen consumption by the different aerobic and endogenous respiration processes........ 97

Table 5.1. Typical characteristics of the influent (diluted swine centrate). 108

Table 5.2. Average NH_4^+-N, NO_2^--N and NO_3^--N concentrations in the influent and effluent of R1 (SRT 7d) and R2 (SRT 11d). Effluent NH_4^+-N and NO_2^--N concentrations were significantly different between phases, for both reactors.

Differences between reactors were not significant (single factor ANOVA 95% confidence interval). 119

Table 6.1. Biomass concentrations based on mass balances (Exp.) in the microalgal-reactor and modelled biomass (Model) using the microalgal-bacterial model. ... 152

Table 6.2. Calibrated parameters for the algal-bacterial model and literature values. 153

Table 6.3 IOA calculated for the modelled parameters in period 1A and 1C..... 158

Table 7.1. Calibrated parameters for the nitrogen storage, phototrophic growth on both nitrogen storage and external ammonium, and autotrophic processes for RT-4, RT-5 and RT-6. 205

Table 7.2. IOA calculated between the modelled data and the measured data for the different compounds for RT-4, RT-5 and RT-6........ 206

Table 8.1. Summary of volumetric and specific ammonium removal rates under the different operational conditions tested in each chapter 216

LIST OF FIGURES

Figure 1.1. Thesis structure and connection among the different chapters 10

Figure 2.1. Microalgae and bacterial oxidation interactions in a microalgal – bacterial consortia. Source: Adapted from Muñoz and Guieysse (2006). OHO: Heterotrophic organisms, PHO: phototrophic organisms, and P: phosphorous.... 16

Figure 2.2. Algae granules containing the algae strains: *Chlorella* sp. and *Phormidium* sp. (Tiron et al. 2017) 18

Figure 2.3.The three most used algal system configurations. A) High rate algae pond, B) Closed tubular photobioreactor, and C) Flat panel airlift reactor (Source: Wang et al., 2018) 26

Figure 2.4. Volumetric productivity of a photobioreactor r^u_x as a function of the biomass concentration C_x. Light intensity at the back of the reactor $I_{ph,PAR}$ (d) and the compensation light intensity $I_{ph,\ PARc}$, are also shown. (Source: Janssen and Lamers 2013) 32

Figure 3.1. Open flat panel reactor (FPR) used in the experiments. 43

Figure 3.2. Evolution of biomass concentrations in FPR1 (A) and FPR2 (B) along the experimental periods. TSS concentration in the FPR (···●···), and VSS concentration in the reactor (··⊙···) 50

Figure 3.3. Biomass production in FPR1 and FPR2 during periods 2 to 7. FPR1 (■), FPR2 (■) 51

Figure 3.4. Concentrations of nitrogen compounds in the effluents of FPR1 (A), and FPR2 (B), along the experimental periods. Legend: ( ) effluent NH_4^+-N, ( ) effluent NO_2^--N, ( ) effluent NO_3^--N, and ( ) influent NH_4^+-N.. 58

Figure 3.5. Total ammonium removal rates for nitrifiers (AOB & NOB) and algae based on the nitrogen balance in FPR1 (A) and FPR2 (B) during the different operational periods. NOB (■), AOB (◫) and algae (▧)...................................... 59

Figure 4.1. Operational scheme, composition of the synthetic wastewater and SRT's applied in the different operational periods assessed in this study. The duration of each phase is presented in minutes below each scheme. Aerobic phase (▦), anoxic phase (□), settling phase (□), and effluent withdrawal phase (■)....................... 77

Figure 4.2. Suspended solids concentration in the sequencing-batch photobioreactor during the 4 periods. Total suspended solids concentration (TSS) (··■··), volatile suspended solids (··□··).. 83

Figure 4.3. Daily nitrogenous concentrations in the reactor along the experimental periods. Influent NH_4^+-N (—◆—), effluent NH_4^+-N (—◇—), effluent NO_3^--N (—●—), and effluent NO_2^--N (—▲—). 87

Figure 4.4. Variation of nitrogen compounds and dissolved oxygen during a SBR cycle scheme for day 47 (Period 1) and day 117 (Period 2B). The trends of the N-compound and oxygen concentrations during period 2B were similar to periods 2A and 2C. Anoxic refers to the dark periods and aerobic to the light periods. NH_4^+-N (—◆—), NO_3^--N (—●—), and NO_2^-N (—▲—)... 88

Figure 4.5. Biomass composition at the different SRTs tested. VSS nitrifiers (▧), VSS OHO (▩),VSS Algae (■). 944

Figure 5.1. Operational steps of the PSBRs during one cycle of a) Phase 1: no sodium acetate addition and of b) Phase 2: with sodium acetate addition at the start of the dark period. 109

Figure 5.2. Schematic of model structure elaboration, combining modified ASM3 and new algal processes to propose the algal-bacterial model. 115

Figure 5.3. Influent and effluent ammonium nitrogen (NH_4^+-N), nitrite nitrogen (NO_2^--N) and nitrate nitrogen (NO_3^--N) concentrations over time in R1 (SRT 7d) and R2 (SRT 11d).. 118

Figure 5.4. Light intensities measured at varying distance from light source inside the PSBR, and varying TSS concentrations (C1-C7).. 120

Figure 5.5. Estimation of irradiated zones at varying light intensities (R1: 98%, R2: 75% of reactor volume), and completely dark zones (R1: 2%, R2: 25% of reactor volume) inside both PSBRs. 1222

Figure 5.6. Profiles of model predictions and experimental data of nitrogen species and DO for both reactors, during one cycle (Phase 2, Day 49)........................1255

Figure 5.7. Simulations of the base model (uncalibrated) considering an algal system in R1, i.e. with no bacterial processes incorporated..1277

Figure 6.1. Graphical scheme of the conceptual arrangement of tanks of the photo-activated bioreactor in Aquasim. 139

Figure 6.2. Sensitivity analysis of the mathematical prediction of (A) NH_4^+-N, (B) NO_2^--N, (C) NO_3^--N, (D) COD and (E) O_2, with respect to: $\boldsymbol{Is}$ (.........), $\boldsymbol{k}$ (----·), $\boldsymbol{\mu m, AOB}$ (— · ·), $\boldsymbol{\mu m, NOB}$ (— · -), $\boldsymbol{\mu m, H}$ (— —), $\boldsymbol{\mu m, P}$ (——), and $\boldsymbol{kSTO}$ (══)...1466

Figure 6.3. Modelled and measured data for NO_3^--N (A), NH_4^+-N (B), COD (C) and O_2 (D) concentrations during period 1B. Solid line: model data during period 1B; measured data during period 1B in cycles C1 (□), C2 (◇), C3 (○), C4 (△) and C5 (*). 1488

Figure 6.4. Comparison between the modelled and measured biomass in the microalgal-bacterial reactor for period 1B. X_T: Total biomass, X_P: Phototrophic biomass, X_H: Heterotrophic biomass, X_AOB+X_XNOB: Ammonium and nitrite oxidising bacteria. 15050

Figure 6.5. Calibrated N-compounds concentrations for period 1B with a SRT of 26 days: NH_4^+-N (▬), NO_3^--N (▬) and NO_2^--N (▬) modelled concentration and NH_4^+-N (●), NO_3^--N (■) and NO_2^--N (△) measured concentration. The grey-shaded areas correspond to the dark phases (lights turned off) during the cycles. 1566

Figure 6.6. Description of the concentrations of ammonium (▬), nitrate (▬), nitrite (▬), oxygen (▬), COD (▬) for period 1B with a SRT of 26 days after calibration and comparison with the measured data (-o-). 1577

Figure 6.7. Validation of the model showing the concentrations of the N-compounds for period 1A (A) and 1C (B) with a SRT of 52 days and 17 days, respectively: NH_4^+-N (▬), NO_3^--N (▬) and NO_2^--N (▬) concentrations. The grey-shaded areas correspond to the dark phases (lights turned off) during the cycles. 1599

Figure 6.8. Prediction of the N-compounds and oxygen concentration for shorter SRTs: 10, 5 and 0.9 day. NH_4^+-N (▬), NO_3^--N (▬), NO_2^--N (▬), and O_2 (▬) concentrations. The gray-shaded areas correspond to the dark phases (lights turned off) during the cycles. 1677

Figure 7.1. Flat panel reactor used as parent microalgal-bacterial reactor, the light was applied perpendicular to the largest cross-sectional area............................1799

Figure 7.2. Respirometric unit used to perform the respirometric tests. (A) Reactor of 1 L connected to the double wall heated respirometer vessel and (B) Close up of the respirometric vessel, in which the oxymeter is placed...............................18080

Figure 7.3. Suspended solids concentrations during the entire operation of the microalgal-bacterial reactor. TSS in the reactor (●), VSS in the reactor (○), effluent TSS (◆), and effluent VSS (◇)...1877

Figure 7.4. Light measurements (µmol m^{-2} $s^{-1)}$ in the flat panel reactor at C1: 1.56 (± 0.18) gTSS L^{-1} and C9 corresponds to zero as it corresponds to the synthetic medium fed to the reactors. 1888

Figure 7.5. Nitrogen compounds concentrations during the entire operation of the microalgal-bacterial reactor. Influent NH_4^+-N (●), effluent NO_2^--N (◇), effluent NO_3^--N (■), and effluent NH_4^+-N (○). 19191

Figure 7.6. RT-1 with initial ammonium concentration of 13.4 mg NH_4^+-N L^{-1}. NH_4^+-N (●), NO_2^--N (◇), NO_3^--N (■), and O_2 (▬). 1922

Figure 7.7. RT-2 with initial ammonium concentration of 13.4 mg NH_4^+-N L^{-1} and addition of ATU to stop nitrification and nitritation. NH_4^+-N L^{-1} concentration (●) and O_2 concentration (▬)...1955

Figure 7.8. RT-3 with initial ammonium concentration of 0.22 mg NH_4^+-N L^{-1} and addition of ATU to stop nitrification and nitritation. NH_4^+-N L^{-1}. NH_4^+-N (●), NO_2^--N (◇), NO_3^--N (■), and O_2 (▬)..1988

Figure 7.9. Results of the RT-4 data measured for calibration of the expanded model. NH_4^+-N (●), NO_2^--N (◇), NO_3^--N (■), and O_2 (▬)....................................2022

Figure 7.10. Results of RT-5 (A) and RT-6 (B) measured for calibration of the expanded model. NH_4^+-N (●), NO_2^--N (◇), NO_3^--N (■), and O_2 (▬)................2033

Figure 7.11. Calibration results for RT-4. Measured NH_4^+-N (●), measured NO_2^--N (◇), measured NO_3^--N (■), measured O_2 (△), modelled NH_4^+-N (▬), modelled NO_2^--N (▬), modelled NO_3^--N (▬), and modelled O_2 (▬)..................................2066

Figure 7.12. Calibration results for RT-5 and RT-6. Measured NH_4^+-N (●), measured NO_2-N (◇), measured NO_3^--N (■), measured O_2 (△), modelled NH_4^+-N (▬), modelled NO_2^--N (▬), modelled NO_3^--N (▬), and modelled O_2 (▬)..2077

Figure 8.1. Key important findings of the research on algal-bacterial systems performed in this PhD study. 2144

Figure 8.2. Scheme of the proposed holistic approach for treatment of domestic, industrial and agricultural wastes. CHP: combined heat and power system, N: nitrogen and P: phosphorous...22816

About the author

Angelica Rada was born in Santa Marta, Colombia. She is a Civil Engineer (2008) form the Universidad Nacional de Colombia (Bogota). She worked as civil engineer in Colombia in different projects during 2008 and 2010 in the area of hydraulics and environmental technology. In 2012, she successfully completed her MSc. studies on sanitary engineering and municipal water and infrastructure. Angélica enrolled as PhD fellow between 2013 - 2018 in the Environmental Engineering & Water Technology department at UNESCO - IHE, Delft, and Wageningen University. Her PhD research focused on the treatment of ammonium rich wastewater using microalgal-bacterial consortia. Over the past nine years, she has worked with a team of experts on wastewater treatment, environmental technologies, sanitation techniques and civil engineering activities in Colombia and The Netherlands inside and outside academia. This has helped her to develop different abilities on a technical and personal level as a civil and sanitary engineering working in the field of environmental biotechnology.

Journals publications

Rada-Ariza, María, A., Lopez-Vazquez, C.M., Van der Steen, N.P., Lens, P.N.L.,. Ammonium removal mechanisms in a microalgal-bacterial sequencing-batch photobioreactor at different SRTs. Algal Research (Major revision).

Rada-Ariza, María, A., Lopez-Vazquez, C.M., Van der Steen, N.P., Lens, P.N.L., 2017. Nitrification by microalgal-bacterial consortia for ammonium removal in flat panel sequencing batch photo-bioreactors. Bioresource Technology. https://doi.org/10.1016/j.biortech.2017.08.019.

Arashiro, L.T., Rada-Ariza, A.M., Wang, M., Steen, P. van der, Ergas, S.J., 2016. Modelling shortcut nitrogen removal from wastewater using an algal-bacterial consortium. Water Science and Technology 782–792. https://doi.org/10.2166/wst.2016.561.

Van der Steen, P., Rahsilawati, K., Rada-Ariza, A.M., Lopez-Vazquez, C.M., Lens, P.N.L., 2015. A new photo-activated sludge system for nitrification by an algal-bacterial consortium in a photo-bioreactor with biomass recycle. Water Sci. Technol. 72, 443–450. https://doi.org/10.2166/wst.2015.205.

Conference proceedings

Zalivina N, Keeley R, Arashiro LT, Rada-Ariza A.M., Scott K., Van der Steen P., Ergas S.J. (2017) Effect of Solids Retention Time on Nitrogen Removal and Microbial Consortium in a Novel Algal-Bacterial Shortcut Nitrogen Removal System. Proc Water Environ Fed 2017:225–230. doi: 10.2175/193864717821494213.

Rada-Ariza, A.M., Lopez-Vazquez C.M, Van der Steen N.P., Lens P.N.L (2017) Respirometric method for determination of microalgae and microalgae-bacteria consortia and microalgae consortia growth kinetics. In: 1st IWA Conference on Algal Technologies for Wastewater Treatment and Resource Recovery, Delft, The Netherlands.

Arashiro, L.T., Rada-Ariza, A.M., Wang, M., Steen, P. van der, Ergas, S.J. (2016) Modelling shortcut nitrogen removal from wastewater using an algal-bacterial consortium. In: The 13th IWA Leading Edge Technology conference (LET) on Water and Wastewater technologies, IWA Pub., Jerez de la Frontera, Spain.

Rada-Ariza AM, Alfonso-Martinez A, Leshem U, Lopez-Vazquez, C.M., Van der Steen, N.P., Lens, P.N.L. (2016) Nitrification in Pilot-Scale High Rate Algae Ponds Operated as Sequencing Batch Photo-bioreactors for Anaerobically Treated Domestic Wastewater. In: Innovations in Pond Technology for Achieving Sustainable Wastewater Treatment. IWA Pub., Leeds, U.K.

Rada-Ariza A., Rahman A, Zalivina N, Lopez-Vazquez, C.M., Van der Steen, N.P., Lens, P.N.L. (2015) Sludge Retention Time Effects on Ammonium Removal in a Photo-CSTR Using a Microalgae-Bacteria Consortium. IWA Pub., Gdansk, Poland.

Rada-Ariza A.M., Torres A., (2011) Preliminary Assessment of the Influence of Salitre Basin Rainfall on the Pumping Operation of Salitre Wastewater Treatment Plant (Bogota). In: the 12th International Conference on Urban Drainage, Porto Alegre, Brazil.

The research described in this thesis was financially supported by the Colombian organization: Departamento Administrativo de Ciencia, Tecnología e Innovación, COLCIENCIAS. In addition, the financial support by IHE-Institute for Water Education towards the end of the P.hD.

The author would like to acknowledge the design of the cover by Marvin Stiefelhagen.

Netherlands Research School for the
Socio-Economic and Natural Sciences of the Environment

D I P L O M A

For specialised PhD training

The Netherlands Research School for the
Socio-Economic and Natural Sciences of the Environment
(SENSE) declares that

Angélica Rada Ariza

born on 12 December 1985 in Santa Marta, Colombia

has successfully fulfilled all requirements of the
Educational Programme of SENSE.

Delft, 16 November 2018

The Chairman of the SENSE board

Prof. dr. Martin Wassen

the SENSE Director of Education

Dr. Ad van Dommelen

The SENSE Research School has been accredited by the Royal Netherlands Academy of Arts and Sciences (KNAW)

KONINKLIJKE NEDERLANDSE
AKADEMIE VAN WETENSCHAPPEN

The SENSE Research School declares that **Angélica Rada Ariza** has successfully fulfilled all requirements of the Educational PhD Programme of SENSE with a work load of 36.1 EC, including the following activities:

SENSE PhD Courses

- Environmental research in context (2013)
- Research in context activity: 'Co-organizing 1st IWA Conference on Algal Technologies for Wastewater Treatment and Resource Recovery (16-17 March 2017, Delft)'

Other PhD and Advanced MSc Courses

- Laboratory training course, IHE-Delft (2013)
- Microalgae biotechnology, Wageningen University (2014)
- Scientific writing, Premier Taaltraining, Utrecht University (2015)
- Advanced scientific writing, Premier Taaltraining, Utrecht University (2017)

Management and Didactic Skills Training

- Supervising three MSc students with theses (2013-2016)
- Monitoring of MSc. Students in microbiology laboratory: Anaerobic treatment session for the urban water and sanitation in IHE-Delft (2015 and 2017)
- Organization of the "MSc day for the students 2015", a workshop for the Environmental Science Programme in IHE-Delft, Institute for Water Education (2015)

Oral Presentations

- *Sludge retention time effects on ammonium removal in a photo-cstr using a microalgae-bacteria consortium*. IWA Nutrient Removal and Recovery 2015: moving innovation into practice. 18-21 May 2015, Gdansk, Poland
- *Nitrification in pilot-scale high rate algae ponds operated as sequencing batch photo-bioreactors for anaerobically treated domestic wastewater*. IWA 11th Specialist Group Conference on Wastewater Pond Technologies – Innovations in Pond Technology for Achieving Sustainable Wastewater Treatment, 21-23 march 2016, Leeds, United Kingdom
- *Modelling shortcut nitrogen removal from wastewater using an algal-bacterial consortium.* The 13TH IWA Leading Edge Conference on Water and Wastewater, 13-16 June 2016, Jerez de la Frontera, Spain
- *Respirometric method for determination of microalgae and microalgae-bacteria consortia growth kinetics.* 1st IWA Conference on Algal Technologies for Wastewater Treatment and Resource Recovery, 16-17 March 2017, Delft, The Netherlands

SENSE Coordinator PhD Education

Dr. Peter Vermeulen

For Product Safety Concerns and Information please contact our EU representative GPSR@taylorandfrancis.com Taylor & Francis Verlag GmbH, Kaufingerstraße 24, 80331 München, Germany

Printed and bound by CPI Group (UK) Ltd, Croydon, CR0 4YY
01/05/2025
01858618-0003